Proceedings of the

Dirac Centennial Symposium

Proceedings of the Dirac Centennial Symposium

Florida State University, Tallahassee, USA

6 – 7 December 2002

edited by

Howard Baer & Alexander Belyaev

Florida State University, Tallahassee, USA

World Scientific

NEW JERSEY • LONDON • SINGAPORE • SHANGHAI • HONG KONG • TAIPEI • BANGALORE

Published by

World Scientific Publishing Co. Pte. Ltd.
5 Toh Tuck Link, Singapore 596224
USA office: Suite 202, 1060 Main Street, River Edge, NJ 07661
UK office: 57 Shelton Street, Covent Garden, London WC2H 9HE

British Library Cataloguing-in-Publication Data
A catalogue record for this book is available from the British Library.

PROCEEDINGS OF THE DIRAC CENTENNIAL SYMPOSIUM

ISBN 981-238-412-X

Printed in Singapore by World Scientific Printers (S) Pte Ltd

1) Stanley Deser
2) Pierre Ramond
3) Paul Langacker
4) Brian Serot
5) Laurie Brown
6) Monica Dirac
7) Leopold Halpern
8) Joe Polchinski
9) Howie Baer
10) Vernon Barger
11) Elihu Abrahams
12) Vasken Hagopian
13) Jonathan Bagger
14) Heaya Summy
15) Andres Rodriguez
16) Muslema Pervin
17) Sharon Schwerzel
18) Sinjini Sengupta
19) Susan Blessing
20) Suharyo Sumowidagdo
21) Roman Jackiw
22) William Marciano
23) Harrison Prosper
24) Zohreh Parsa
25) Bonnie Todd
26) Isaac Rytel
27) Doan Nguyeu
28) Vernon Stokes
29) Layla Hormozi
30) Alvin Kiswandhi
31) Jorge O'Farrill
32) Jubri Taruna
33) Chew-Lean Lee
34) George DeRise
35) Xerxes Tata
36) Sharon Hagopian
37) Csaba Balazs
38) Harry Ringermacher
39) Richard Bardo
40) Jeff Owens
41) Eric Jobillong
42) Jorge Piekarewicz
43) Genaro Toledo
44) Mark Riley
45) Cengiz Sen
46) Karah Manning
47) David Mozzoni
48) Martin Kretzschmar
49) Chris Jackson
50) Scott Riggs
51) Lucas Blaszczyk
52) Kunnat Sebastian
53) Ali Nayeri
54) Tuan Tran
55) Chenggany Tao
56) Ashwani Kumar
57) Cosimo Felline
58) Paul Eugenio
59) Gabe Pendas
60) Konstantin Matchev
61) Laura Reina
62) Daniel Auto
63) Azar Mustafayev
64) Albert Wynn
65) Alexander Belyaev
66) Tadas Krupovnickas
67) Tjipto Juwono
68) Harvey Goldman
69) Bruce Ferguson
70) Fernando Febres-Cordero
71) Bernd Berg
72) Shixiang Feng
73) Todd Adams
74) Martin Smith
75) Bill Green
76) Nicholas Huebner
77) Ivo Janier Calderin
78) Simon Capstick
79) Kurtis Johnson
80) Volkan Sevim
81) Xiaohang Zhang
82) Sangjin Lee
83) Alexander Velytsky
84) Kyoungchul Kong
85) Hugh Logan
86) Aseshkrishna Datta

CONTENTS

PREFACE

Paul Adrian Maurice Dirac is one of the icons of modern physics. Born in 1902, he began formal work on the new Quantum Theory beginning in 1925. Dirac's work provided the mathematical foundations of quantum mechanics. He also made key contributions to quantum field theory and quantum statistical mechanics. He is perhaps best known for formulating the Dirac equation, a relativistic wave equation which described the spin and magnetic properties of the electron, and also predicted the existence of anti-matter.

Dirac retired from his position as Lucasian Professor at Cambridge University in 1969. He joined the faculty at Florida State University in Tallahassee, Florida in 1970, where he remained until his death in 1984.

The Dirac Centennial Symposium was a two day gathering of eminent researchers, faculty and students on December 6-7, 2002 to commemorate the contributions of Professor Dirac to all areas of physics, and to assess their impact upon frontier research. After an introductory overview, this volume contains contributions from Laurie Brown (Northwestern University), Leopold Halpern (FSU), Pierre Ramond (UF), Brian Serot (Indiana), Frank Wilczek (MIT), Maurice Goldhaber (BNL), Bill Marciano (BNL) Paul Langacker (Penn), Vernon Barger (Wisconsin), Roman Jackiw (MIT), Stanley Deser (Brandeis) and Joe Polchinski (UCSB). A special contribution from Dirac's daughter, Monica Dirac, presents a portrait of Paul Dirac, father and family man. In addition, presentations were made by Elihu Abrahams (Rutgers), Jon Bagger (Johns-Hopkins), Joe Lykken (Fermilab) and Andre Linde (Stanford), although these contributors were unable to provide written manuscripts.

We would like to take the opportunity to thank the Dirac symposium organizing committee, Csaba Balazs, Bill Green, Vasken Hagopian, Leopold Halpern, Jeff Owens, Jorge Piekarewicz, Harrison Prosper, Laura Reina, Bob Schrieffer and Bonnie Todd for all their help. We thank Sharon Schwerzel and the staff of the Dirac Science Library for assembling a wonderful display of Dirac memorabilia, which everyone enjoyed. We thank Sherry Beasley, Kathy Mork and Kristie Johnson for making the symposium run smoothly, and putting in long hours. Thanks to Ken Ford and Scott Baxter for making posters and displays, and taking pictures. We thank Yazid Johnson and Paul Harvey for the banquet music. From FSU, we thank Department of Physics Chair Kirby Kemper, Provost Larry Abel, Vice-President of Research Ray Bye and Dean of Arts and Science Don Foss for their financial support. We also thank the US Department of Energy and the National Science Foundation for their financial support for the FSU Dirac Centennial Symposium.

This material is based upon work supported by the National Science Foundation under Grant. No. 0225593.

Howard Baer and Alexander Belyaev

Proceedings of the Dirac Centennial Symposium
Howard Baer and Alexander Belyaev

Introduction

Howard Baer
Florida State University
Department of Physics
Tallahassee, FL 32306, USA
E-mail: baer@hep.fsu.edu

The year 2002 marked the 100th anniversary of the birth of Paul Adrian Maurice Dirac, one of the founding fathers of modern physics, and faculty member at Florida State University from 1970 until his death in 1984. It is just cause for organizing a symposium to re-examine Dirac's work, and the considerable impact it has had on forefront research in almost all branches of physics. Thus, a two day symposium, the Dirac Centennial Symposium, was organized by the Florida State University Department of Physics, and was held on December 6-7, 2002. In fact, the FSU Dirac symposium was one of *four* such symposia held in 2002. They include:

- Beauty in Physics: the Life and Work of Paul Dirac, Institute of Physics, Brighton Centre, UK, April 9, 2002,
- Dirac Centennial Celebration, Cambridge University, July 20, 2002,
- Dirac Centenary Conference, Baylor University, Sept. 30- Oct. 2, 2002,
- Dirac Centennial Symposium, Florida State University, Dec. 6-7, 2002.

The Dirac Centennial Symposium is not the first Dirac Symposium to be held at Florida State University. In 1977, a 75th birthday/50th anniversary of Dirac equation symposium was held.[1] Speakers included Gerard 't Hooft,

John Kogut, Frank Wilczek, J. J. Sakurai, John Ellis, David Politzer, Ken Johnson, Yuval Ne'eman, Claudio Tietelboim, Leopold Halpern, Sigfried Wouthuysen, Ken Wilson, Freeman Dyson, Eugene Wigner (as the summary speaker), and of course Dirac himself. Dirac's contribution was entitled *Consequences of varying G*, and addresses his large numbers hypothesis and time variation of physical constants.

Dirac was born on August 8, 1902 in Bristol, UK to Charles Dirac and Francis Holton. He had an older brother, Reginald, and a younger sister, Beatrice. Dirac noted later in life that his upbringing under his father's strict regime had a strong effect on him, giving rise to his reticent personality, and his relations with his father were always strained.

Paul Dirac entered University of Bristol to study electrical engineering, but went on to postgraduate work in mathematics and physics. In 1925, Dirac's graduate advisor brought some work by Heisenberg on matrix mechanics to his attention. Dirac soon found a bridge between the Poisson brackets of classical mechanics and the commutators of quantum theory, allowing him to devise a scheme for passage from classical to quantum theory in general. In May, 1926, he received his doctorate degree in physics. In the next two years, Paul Dirac laid many of the foundations of quantum mechanics, especially his transformation theory, which he always referred to as his "darling". Then in 1927 he found the relativistic wave equation which bears his name, and which gave a fundamental description of spin and magnetic moments for spin-$\frac{1}{2}$ particles. Dirac grappled with the problem of negative energy solutions for several years before proposing his "Dirac sea", and the concommitant prediction of the existence of the positron. He was awarded the Nobel prize in 1933 along with Erwin Schroedinger. In 1937, he was married to Margit Wigner.

1933 also ends what Abraham Pais refers to as Dirac's "heroic period", wherein he made an astonishing variety of fundamental contributions to quantum theory.[2] However, Dirac was by no means inactive after this period. As remarked by Polchinski in this volume, Dirac made key advances in theory of magnetic monopoles, path integrals, light cone dynamics, membrane actions, conformal and de Sitter symmetries, constrained Hamiltonian dynamics and canonical formulation of gravity. "For an anticlimax, that is a pretty good career."

Much as Einstein stepped out of the limelight in his later years, Dirac pursued his own directions in the latter part of his career. In part, this was because he regarded the ultimate formulation of renormalization in quantum electrodynamics, and quantum field theory in general, as ugly, inadequate and probably wrong. He focused much attention on alternative formulations of QED, to the neglect of understanding the other particles and interactions which absorbed the attention of much of the physics research community.

Upon retirement from his chair as Lucasian Professor at Cambridge University in 1969, several universities in the US pursued Dirac as a faculty member. Urged on by the late Joe Lannutti, he chose to come to Florida State University in Tallahassee in part for his wife Margit's sake (she wanted the warm weather), and partly for his own sake, because Tallahassee offered opportunities for long walks, and swims in nearby lakes, rivers and springs, which Dirac enjoyed. Joe Lannutti remarked in a letter to Abraham Pais[2] that

> Dirac was most happy in Tallahassee, he really changed. In Cambridge, he only went to the University for classes and seminars but otherwise worked at home. In Tallahassee he came diligently all day, ate lunch with the boys, took a nap after lunch. His wife would pick him up in the late afternoon... We treated him like one of the boys...did not indulge in much red carpet treatment. He liked that.

Pais goes on to note that

> Dirac's writings in the Florida period are simply prolific. He published over 60 papers in those last 12 years of his life, most of them reviews of past events, including a short book on General Relativity.

Dirac passed away on October 20, 1984, and is buried on Tallahassee's north side, in Roselawn Cemetary. Margit passed away in summer of 2002, and is buried next to Paul.

Our goals in organizing the FSU Dirac Centennial Symposium included 1. examining Dirac's life and work from an historical and personal level, and 2. examining Dirac's continuing impact on frontier research areas. In this respect, we focused on the three areas of physics research which are active at FSU: high energy physics, nuclear physics and condensed matter physics. Finally, 3. we wanted to survey the developments in areas of physics that were of special interest to Paul Dirac. Unfortunately, a snowstorm hit the east coast of the USA on the day before the symposium, and two of our speakers– Frank Wilczek and Maurice Goldhaber– were unable to attend. Nevertheless, they both sent in their contributions to this volume.

Laurie Brown (Northwestern University) spoke on the topic of *Paul Dirac: Building Bridges of the Mind*, and illustrated how much of Dirac's work centered on reconciling apparently disparate themes in physics: wave mechanics and matrix mechanics, relativity and quantum mechanics. Leopold Halpern (FSU), our next speaker, was Dirac's personal assistant and a research scientist at FSU. He presented his fascinating reminiscences of Dirac, and also explains some of the research directions he pursued, which were influenced by Dirac. Monica Dirac, the older of Paul's and Margit's children, presented the banquet speech, a wonderful collection of stories about Paul Dirac, family man and loving father.

Frank Wilczek (MIT) contributed to this volume his fine essay *The Dirac Equation*. The Dirac equation is exceptional in part because it emerged from the requirements of beauty and symmetry, rather than being directly motivated by experimental data. This philosophy, that the fundamental equations of physics be above all beautiful, is a central theme and guiding principle in Dirac's life and work. Dirac's equation is also notable in the evolution of its interpretation. Originally, Dirac interpreted it as a wave equation, acting on a wave function. This interpretation led to the problem of negative energy states, and to Dirac's postulation of the vacuum being a *Dirac sea* filled with negative energy electrons. The stability of the negative energy "sea" was ensured by the Pauli exclusion principle. In its day, many of Dirac's contemporaries, notably Pauli himself, found this untenable, in spite of Dirac's successful prediction of the existence of the positron. Nowadays, the fundamental utility of the Dirac equation is recognized by re-interpreting it as the equation of motion governing the free relativistic spin-1/2 (Dirac) fermion *field* in quantum field theory. The negative "frequency" solutions correspond to the energy required to create an electron out of the vacuum. Particles and anti-particles appear as excitations of the vacuum, and both are necessary to maintain causality in relativistic quantum field theory. While Dirac himself pioneered much of quantum field theory, and the quantization of the electromagnetic field, it is an enigma that he never accepted the ultimate formulation of quantum electrodynamics, and its later generalization to the Standard Model, even in spite of the astonishing predictivity of the theory.

One of the first astonishing implications of the Dirac equation was that the Landé g factor of the magnetic moment of the electron turns out to have the value $g = 2$, as known by experiment in the 1920s. The magnetic moment of the electron later played a central role in establishing the validity of quantum electrodynamics. Today, the agreement between experiment and theory is good to 8 decimal places! The magnetic moment of the electron's heavier copy, the muon, is still a forfront issue due to recent ultra-precise measurements by the Brookhaven $g-2$ experiment, E821.[3] Bill Marciano (BNL) gives an overview of the developments in $g-2$ from Dirac up to the present time, and how measurements of this quantity have in the past indicated new physics, and even now may be pointing to physics beyond the Standard Model.

Dirac's equation was generalized and incorporated into string theory by Pierre Ramond, of our neighboring University of Florida at Gainesville. Pierre presents *Dirac's footsteps and supersymmetry.* The Dirac equation contains in it the seeds of supersymmetry. Like the Dirac equation itself, supersymmetry originated as a beautiful concept in physics, that was later found to have many wonderful applications. Also, Dirac's formulation of light cone dynamics helped to establish the finiteness of $N = 4$ supersymmetric Yang-Mills quantum field theory.

Stanley Deser (Brandeis) was a good friend of Dirac's in his later years. He presents an essay on *Dirac and General Relativity.* Amongst the founders of quantum mechanics, Dirac was exceptional in that he engaged in meaningful research in GR as well. His Hamiltonian formulation of gravity was in many respects a precursor to later formulations of supergravity.

The influence of Dirac upon condensed matter physics, nuclear physics and high energy physics was examined by Elihu Abrahams (Rutgers), Brian Serot (Indiana) and Jonathan Bagger (Johns-Hopkins), respectively.

A topic of contemporary interest is the nature of neutrinos. Recent decisive measurements of neutrino oscillations show that neutrinos are indeed massive, but are they Dirac or Majorana particles? What are their mixing properties, and are there sources of CP violation in the neutrino sector? These and other questions were examined by Joe Lykken (FNAL) and Vernon Barger (Wisconsin) at the Dirac symposium, and a review is given in this volume by Barger.

As mentioned earlier, Dirac was guided very much by aesthetics in his research in theoretical physics, and not so much by data. This is especially apparent in his modification of Maxwell's equations to allow for the presence of magnetic monopoles. While no meaningful evidence has been established for the existence of magnetic monopoles, nonetheless the theoretical examination of their properties has been a major theme in particle theory, and has spawned a number of search experiments. Indeed, monopoles are to be expected in many grand unified and string theories, where they are expected to have been produced in the very early universe. One of the assets of inflationary cosmology is that the density of relic monopoles from the early universe is expected to be highly diluted, and essentially unobservable today. In these proceedings, Roman Jackiw (MIT) revisits Dirac's monopole, and a derivation of quantization conditions is presented, without reference to vector potentials. Joe Polchinski (UCSB) addresses *Monopoles, Duality and String Theory.* While Dirac showed that the existence of monopoles implies charge quantization, Polchinski suggests the converse, and illustrates it with a number of well-motivated examples.

Dirac was intrigued by the idea that many of the physical constants in nature may in fact vary significantly over cosmological time scales. He made use of this to try to explain many large or small number combinations that arise in physics. In the recent few years, in fact, a group has reported some evidence for time variation of the fine structure constant.[4] In theories with extra dimensions, such as string theories, fundamental constants are related to moduli fields, and are expected to vary with time. Paul Langacker (Penn) points out that various fundamental constants are expected to vary in a correlated fashion. Uncovering the pattern of time variation of physical constants may help determine the underlying (string) theory of the universe.

Maurice Goldhaber (BNL) has had a distinguished career in experimental physics, and was a student of Dirac's at one time. In his essay, he reflects on *Amending the Standard Model of Particle Physics.* Based on the masses and interactions of the fundamental fermions, Goldhaber extracts some patterns, or "rules", which may indicate the presence of new fermions, and/or new interactions.

References

1. *Current Trends in the Theory of Fields: A Symposium in honor of P. A. M. Dirac*, J. E. Lannutti and P. K. Williams, ed. (American Institute of Physics, 1978).
2. A. Pais, in *Paul Dirac: The Man and His Work*, P. Goddard, ed. (Cambridge University Press, 1998).
3. G. Bennett *et al.* (E821 Collaboration), Phys. Rev. Lett. **89**, 101804 (2002).
4. J. K. Webb *et al.*, Phys. Rev. Lett. **87**, 091301 (2001).

Laurie M. Brown

Proceedings of the Dirac Centennial Symposium
Howard Baer and Alexander Belyaev

Paul Dirac: Building Bridges of the Mind

Laurie M. Brown
Department of Physics and Astronomy
Northwestern University
Evanston, Illinois 60201, USA
E-mail: lbrown@northwestern.edu

Paul Dirac was a brilliant and original thinker. He used his physical intuition and his ideal of mathematical beauty to construct bridges between major areas of physics. This article discusses several such important works, including the bridge between quantum mechanics and relativity that led to his prediction of the existence of antimatter.

1. Sketch of Dirac's Early Life

On this historic occasion we are celebrating the centennial of the birth of Paul Dirac at the place he chose to spend the last fourteen years of his life. My task is to give a brief sketch of Dirac's life and work. For more details, the reader should consult the bibliography at the end of this article. Physicists know Paul Dirac to be one of the intellectual giants of all time, but his name is little known to the general public. On 13 November 1995 a plaque honoring Dirac was installed in Westminster Abbey near the grave of Isaac Newton. On that occasion, Stephen Hawking, the current Lucasian Professor at Cambridge University delivered an address. Here is some of what he said:[a]

Paul Adrian Maurice Dirac... went on to become the Lucasian Professor at Cambridge and to win a Nobel Prize, but was never well known to the public. His death in 1984 drew a short obituary in the *Times*, but otherwise it went almost unnoticed. It has taken 11 years for the nation to recognize

[a]Pais *et al.*, p. xiii.

that he was probably the greatest British theoretical physicist since Newton, and belatedly to erect a plaque to him in Westminster Abbey.

Dirac was born on 8 August 1902 in Bristol, England, and received his early education there. He studied at the Merchant Venturers' Technical College in Bristol, where his father, Charles Dirac, taught French. After earning a degree in electrical engineering at the University of Bristol in 1921 and studying mathematics there for two years, he applied for and won a scholarship at Cambridge and became a graduate student at St. John's College, beginning research in theoretical physics under Professor R.H. Fowler. In 1926 he obtained his doctorate with a thesis entitled "Quantum Mechanics". That is a sketch of how Dirac began his scientific career. What kind of physicist did he become? According to Richard Dalitz "P.A.M. Dirac was Britain's outstanding theoretical physicist in the twentieth century, and certainly one of the world's great physicists over all time."

There are some interesting parallels between Dirac and Newton. Dirac was 82 when he died on October 20, 1984; Newton died in 1727 at age 84. Newton was a prodigy of 24 in 1666, his annus mirabilis, when he made his first major discoveries. Dirac's was 23 in 1925, when he invented the transformation theory of quantum mechanics. Recognizing his genius, Newton's professor, Isaac Barrow, resigned the Lucasian Chair of Mathematics at Cambridge so that Newton could acquire it at age 26; Dirac was elected to that chair in 1932 when he was all of thirty! (The following year he shared the Nobel Prize with Erwin Schrödinger.) Newton's greatest work was called "Principles of Natural Philosophy", the Principia, while Dirac's classic treatise was called "Principles of Quantum Mechanics."

Dirac's and Newton's childhood years have some similarities, although their family circumstances were quite different. Newton, an only child, was born three months after his father's death. When Isaac was three, his mother remarried and moved out, abandoning him to be raised by his grandmother at Woolthorpe. That his stepfather took no interest in Newton became a traumatic episode in his life. According to his biographer Richard Westfall, he became a "sober, silent, thinking lad". Newton never married or had a girl friend, and was a difficult person to deal with.

Paul Dirac had a Swiss father while he grew up, as well as an English mother, a sister, and a brother. However, according to Dirac's biographer, Helge Kragh, "Charles Dirac was a strong-willed man, a domestic tyrant... [who] brought him up in an atmosphere of cold, silence and isolation."[b] In an interview in 1962 [AHQP] Dirac said of his early years, "In those days I didn't speak to anybody unless I was spoken to. I was very much an introvert, and I spent my time thinking about problems in nature."[c] One Dirac story will illustrate his reticence and his dry humor. In the question period after a lecture, a physicist in the audience asked, "Professor Dirac I don't understand how the fourth equation follows from the third." There was silence for some time, until the chair of the session said, "Professor Dirac, will you answer the question?" to which Dirac replied, "It was not a question, it was a comment."

Unlike Newton, Dirac did marry, and raised a family. Although, like Newton, he had no girlfriends, at the age of 35 he married Margit Balasz, a divorced Hungarian lady, who brought a son and a daughter to the marriage. She was the sister of Dirac's friend, the future Nobel prizewinner Eugene Wigner. An oft-repeated anecdote is that when visited by friends shortly after his marriage, they were surprised to find a woman in Dirac's apartment. He said, "I'm sorry—I forgot to introduce you—this lady is Wigner's sister." Paul had two daughters with Margit, and was a caring father. In contrast to Newton, Dirac was kind and gentle, had a good sense of humor, and could be a warm friend. Unlike Newton, Dirac was never interested in alchemy, biblical chronology, or religion.

2. Classical and Modern Physics

It is now time to discuss the work that Dirac did that puts him in the same class with Newton, Maxwell, and Einstein. As Figure 1 illustrates, Dirac's best work bridged the gaps between classical and modern physics and between relativity and quantum mechanics. It is often stated the last subjects are the two major currents in modern, i.e., twentieth century physics. I would insist on a third major current, namely: atomic and nuclear structure physics including, theoretical chemistry, condensed matter physics, elementary particles, etc.

The scale at which quantum mechanics takes over from classical mechanics is given by the quantum of action h, which Max Planck introduced in 1900 in connection with the theory of blackbody radiation. Einstein's suggestion in 1905 that light consists of directed quanta of energy $h\nu$, where ν is the frequency, was considered a very strange idea by almost everyone

[b] Kragh 1990, p.2.
[c] *ibid.*

CLASSICAL PHYSICS	**MODERN PHYSICS**
(CM) Mechanics	**(QM) Quantum Mechanics**
Newton (Principia 1687)	Planck 1900, Einstein 1905
Lagrange, Hamilton, 1800s	Heisenberg 1925,Schrödinger 1925
(CED) Electrodynamics	**(REL) Relativity**
Maxwell *Treatise on E. and M.* 1873	Einstein 1905 (Special)
Lorentz. 1890	Einstein 1913 (General)

(ATM) Atomic Physics
Einstein 1905
Bohr-Rutherford 1913

Dirac's Main Work was to make the following connections:

CM → **QM** (1925)

CED → **QED** (1927)

QM → **REL** (1928)

QED → **REL** (1933-1984)

Fig. 1. Some of the bridges that Dirac made between major branches of theoretical physics.

(including Planck) for almost two decades. However, in 1913 Niels Bohr found a more acceptable use for h, namely to stabilize and structure Ernest Rutherford's nuclear atom. In Bohr's theory only a restricted number of classical planetary orbits were permitted for the electrons, the size of the allowed orbits being determined by Planck's constant.

Bohr and other physicists extended the theory over the next decade, but when applied to atoms more complex than hydrogen, even to the two-electron atom helium, the theory led to contradiction with experiment. In 1925, Werner Heisenberg, then 23 years old (like Dirac), made an enormous breakthrough in atomic physics. He found a way to calculate frequencies and intensities of atomic spectral lines, for complex atomic systems. When Dirac was asked in 1968 to introduce a lecture by Heisenberg at the International Center for Theoretical Physics in Trieste, Italy, this is what he said:[d]

I have the best of reasons for being an admirer of Werner Heisenberg. He and I were young research students at the same time, about the same

[d]Bethe et al. 1989, p.32.

age, working on the same problem. Heisenberg succeeded where I failed. There was a large mass of spectrographic data accumulated at that time and Heisenberg found out the proper way of handling it. In so doing he started the golden age in theoretical physics, and for a few years after that it was easy for any second rate student to do first rate work.

Heisenberg discovered that transitions from one energy state of the atom to another, leading to the emission of a light quantum, must be represented by quantities related not only to the original state, but also to the state resulting from the emission. Before Heisenberg, the excited atom was regarded as an antenna whose radiation depended only on the state it was in. But if the states are labeled by i and f, for initial and final, we are dealing with an energy E_{if}, not just E_i and E_f separately, and similarly for the other relevant atomic coordinates, such as position and momentum. A mathematical quantity that has a double label like that (and obeys certain multiplication rules) is a matrix, and Heisenberg's approach was called matrix mechanics.

When Heisenberg visited Cambridge in September 1925 he left a proof copy of his forthcoming paper on matrix mechanics with R.H. Fowler, Dirac's research supervisor, who passed it on to Dirac. In a short time the young assistant realized that the important new feature of Heisenberg's work was that multiplying the matrices representing different physical observables like position, velocity, and energy did not in general commute. That is, unlike the situation when multiplying ordinary numbers, the order of multiplication was important. Dirac began to call the new quantities q-numbers (for queer or quantum) and the older physical quantities c-numbers (for commuting or classical). Dirac then realized that each of Heisenberg's q-number relations had a corresponding well-known c-number version that could be found in the textbooks of classical mechanics. Dirac called his method for translating classical equations into quantum ones "transformation theory".

Soon after this the Austrian physicist Erwin Schrödinger found another approach that gave the same results as matrix mechanics, but which was based upon a wave picture for the electron, an approach called wave mechanics. Remarkably, Dirac was able to show that his transformation theory was general enough to include both matrix mechanics and wave mechanics and to demonstrate their equivalence. For their achievements, all three physicists were awarded the Nobel Prize in Physics in 1933. Heisenberg received the 1932 prize, which had not been awarded that year, while Schrödinger and Dirac shared the prize for 1933.

I gave Dirac's appreciation of Heisenberg above. This is what he said about Schrödinger in 1972:[e]

...[Of] all the physicists that I met, I think that Schrödinger was the one that I felt to be most closely similar to myself. I found myself getting into agreement with Schrödinger more readily than with anyone else. I believe the reason for this is that Schrödinger and I both had a very strong appreciation of mathematical beauty, and this appreciation of mathematical beauty dominated all our work. It was a sort of act of faith with us that any equations which describe fundamental laws of nature must have great mathematical beauty in them. It was like a religion with us. It was a very profitable religion to hold, and can be considered as the basis of much of our success.

Dirac made many such statements praising mathematical beauty throughout his life, but I will here make only two remarks about it.[f] First, this appears to be the *only* religion that Dirac adhered to. Second, Dirac's work has led to at least two new branches of mathematics: the theory of distributions, and the theory of the so-called Dirac operator that appears in the Dirac equation.

Based on his transformation theory, Dirac wrote a famous treatise, the *Principles of Quantum Mechanics*, which is often compared with Newton's *Mathematical Principles of Natural Philosophy* or *Principia*. Dirac's *Principles* was published in 1930, went through five editions, and has been translated into many languages. The last edition appeared in 1967and has been reprinted every year since then.

In the preface to the first edition, Dirac emphasized the "vast change" from the classical tradition in which one could "form a mental picture in space and time of the whole scheme". This was no longer the case, since the fundamental laws "control a substratum of which we cannot form a mental picture without introducing irrelevancies". Instead we are obliged to rely on the mathematics of transformations, an abstract and symbolic method. However, although *Principles* was very mathematical, Dirac cautioned:

All the same the mathematics is only a tool and one should learn to hold the physical ideas in one's mind without reference to the mathematical form. In this book I have tried to keep the physics to the forefront, by beginning with an entirely physical chapter and in the later work examining the physical meaning underlying the formalism wherever possible.

---

[e]Weiner 1977, p.36.
[f]Kragh 1990 has an entire chapter on this subject.

3. Quantum Statistics

Besides the transformation theory, between 1926 and 1930 Dirac made other major contributions to quantum theory, including quantum statistics, quantum electrodynamics (QED), and the relativistic theory of the electron. Quantum statistics comes into play when we deal with two or more particles of the same type. Photons and electrons obey two different kinds of statistics: E-B, for Einstein-Bose and F-D, for Fermi-Dirac. Both are quite different from the classical statistics, which assumes that the objects considered are identifiably different. Two or more photons of the same frequency and polarization, traveling in the same direction, are not distinguishable. They obey E-B statistics and tend to occupy the same quantum state. This property of photons is what makes the laser possible. He^4 at low temperature is an E-B liquid, and at sufficiently low temperature is a superfluid. On the other hand electrons obey F-D statistics and obey the *Pauli exclusion principle*, which says: Two electrons can *never* occupy the same quantum state. As a result, when electrons are added to atoms they form new shells. They do not collapse into the lowest energy state, as they would with E-B statistics. Similarly, protons and neutrons obey F-D statistics and form nuclear shells. Electrons in a metal behave as an F-D gas. Obviously this is of the greatest importance for the world (as we know it) to exist. Fermi and Dirac invented and applied the F-D statistics in 1926.[g]

4. The Bridge between Classical and Quantum Electrodynamics

Another great work of Dirac was to make a quantum theory of the electromagnetic field in interaction with electrons in 1927. He first showed that the classical electromagnetic field could be represented by a set of oscillators, and then replaced them by a corresponding set of quantum oscillators, obeying the E-B statistics, using the ideas of transformation theory. Dirac was again emphasizing the classical-quantum connection, this

[g]According to Franco Rasetti: "It is well-known that Dirac developed this type of statistics independently of Fermi... [B]oth of Fermi's publications antedate Dirac's by an appreciable time. Dirac was the first to show that the two types of statistics, now usually designated as Bose-Einstein and Fermi-Dirac, are related to the two possibilities of eigenfunctions of a system being either symmetric or antisymmetric with respect to the exchange of the coordinates of two identical particles." F. Rasetti, *Enrico Fermi: Collected Papers, Volume 1* (1962), p.178.

time CED→QED. In this way he explained the wave-particle paradox as follows:[h]

We may replace the set of harmonic oscillators by a train of waves, each Fourier component of the waves being dynamically equivalent to a simple harmonic oscillator. Thus our Einstein-Bose assembly is dynamically equivalent to a system of waves. This provides us with a complete reconciliation between the corpuscular and wave theories of radiation. We may regard radiation either as an assembly of photons satisfying the Einstein-Bose statistics or as a system of waves, the two points of view being consistent and mathematically equivalent.

5. The Bridge between Quantum Mechanics and Relativity

Probably Dirac's greatest achievement was his relativistic theory of the electron, published in early 1928, which led to his prediction of the existence of the anti-electron or positron, which pointed to the unexpected existence of a whole new world—that of antimatter. Dirac did this by bridging the gap between quantum mechanics and relativity.

The electron's spin and magnetic moment had been inferred from accurate measurements of atomic spectra. Arnold Sommerfeld, Heisenberg's professor in Munich, had obtained a formula giving the hydrogen-atom spectrum to relativistic accuracy, by making a relativistic version of the Bohr atom. However, by 1925 that theory was known to be inadequate, and there was *no* basic theory to explain the electron's spin and magnetic moment. To relate these three things was recognized as a brilliant success of Dirac's new electron theory, and application to basic processes like Compton scattering, i.e., scattering of photons on electrons, brought further successes. However, what led to antimatter was at first considered a failure of the theory, as I shall now explain.

Schrödinger described the electron by a wave function Ψ, a c-number, while Dirac needed four wave functions $\Psi_1\Psi_2$ $\Psi_3\Psi_4$ to describe the electron. (See Figure 3.) Two wave functions are needed to describe an electron that could spin either clockwise or counter clockwise. But the Dirac equation had twice as many solutions. That is because when one starts (as Dirac did) with the equation $\mathrm{E}^2 = \mathrm{p}^2 + \mathrm{m}^2$ and take the square root of both sides one gets both positive and negative energies. Of course, the same thing happens

[h]Dirac 1930, p.223.

RELATIVITY AND QUANTUM MECHANICS

$E^2 = m^2c^4 + p^2c^2, \quad p = mv$

(with c=1) $\quad E^2 = m^2 + p^2$ Eq. R-C

For small p, $E \to m$, and $E + m \approx 2m$

Thus: $\quad E - m = p^2/2m$ Eq. NR-C

When translated into the language of quantum mechanics:

$(E - m)\Psi = (p^2/2m)\Psi$, Eq. NR-Q

E and p are now q-numbers. This is Schrödinger's equation.

For the relativistic Schrödinger's equation,use Eq. R-C:

$E^2\Psi = (m^2 + p^2)\Psi$ Eq.R-Q

No agreement with experiment found for electrons!

However, Dirac's transformation theory requires a linear equation like this:

$E\Psi = (\beta m + \alpha p)\Psi$ Eq.D

The momentum p has really three components, p_x, p_y,p_z

so $\alpha p = \alpha_x p_x + \alpha_y p_y + \alpha_z p_z$.

***To satisfy relativity**, we must still have*

$E^2 = (\beta m + \alpha \cdot p)^2 = m^2 + p^2$.

This is possible only if the three alphas and β *are not numbers but 4x4 matrices and* Ψ *is a column vector with 4 entries* $\Psi_1\Psi_2\Psi_3\Psi_4$.

At first glance this looks crazy, but the theory produced these spectacular predictions:

1) The electron has a spin: its value is (1/2) (h/2π)

2) The electron is a magnet: its strength is (e/mc)(h/2π)

3) The H-atom energy levels are correctly given to relativistic accuracy

4) Antimatter exists

Fig. 2. Origin of the Dirac Equation.

in classical relativity, but there is a difference.

Classically, there is a range of forbidden energies from +mc^2 to -mc^2. In classical physics, there is no way to pass from positive to negative energies because of this forbidden gap, so if we start out with positive energies, then in a classical world, we would never see negative energies. But in quantum mechanics, the electron can make a quantum jump from a positive to a negative energy, emitting energy 2mc^2 or greater. That is just analogous to the quantum jump in an atom. One cannot simply ignore the negative energy states as they turn out to be essential to obtain the good agreement with experiment.[i] Thus Dirac's theory, in spite of its successes, ran the risk of being ridiculed. Heisenberg called it "the saddest chapter of modern physics."[j]

It took several years to discover what was really going on. At first Dirac thought of what is called the hole theory. He assumed that all the negative energy electron states (the "holes") are already occupied by electrons. Thus they are not available, because of the exclusion principle. But an electron in a hole could obtain enough energy to jump into a positive energy state (e.g., by absorbing a gamma ray). How would such an unfilled hole behave? It must behave like a positive charge, since we have taken away a negative charge. If an electric field were present, it would move, like a bubble in a liquid.

Suppose that some holes are not filled—what could they be? At first, Dirac suggested that they might be protons, even though protons are almost 2000 times heavier than electrons. When others showed (Hermann Weyl, Robert Oppenheimer) that electrons and holes must have the same mass, Dirac made the bold and risky suggestion that they were positive electrons, what we now call positrons, even though no such particles had ever been seen. This prediction, made in May 1931,[k] was experimentally confirmed a year later in cloud chambers exposed to the cosmic rays. Carl Anderson at Caltech saw individual positive electrons, and Blackett and Occhialini in Bristol, England, saw the production of electron-positron pairs, just as the hole theory of Dirac had predicted.

Dirac had also suggested that antiprotons (negative protons) should exist, and these were also found—much later—in 1955. Antimatter itself, that is a hydrogen anti-atom, consisting of an antiproton and a positron, was first observed at CERN in Geneva in 1996. Recently anti-hydrogen has been produced in quantity at CERN. It appears that every particle in nature has an anti-particle (for certain neutral particles, the antiparticle is itself).

---

[i]Even the classical limit of Compton scattering, the well-known Thomson cross-section for electromagnetic scattering from electrons fails if the negative energy contributions are not included.

[j]Heisenberg, letter to Pauli, May3, 1928.

[k]Dirac 1931, p. 61.

6. And Other Bridges

In addition to his important early work and the prediction of antimatter, Dirac had many other prescient ideas throughout his lifetime that have strongly influenced the course of theoretical physics. Some of these are listed in the Appendix and other contributors to this symposium have expanded on them. I only mention that although Dirac was never content with the local quantum field theory that he initiated and never accepted renormalization as more than a temporary expedient, nevertheless many of the more recent developments, including *QED*, the Standard Model, and quantum gravity, are based upon his pioneering papers.

Appendix. Some Other Important Work of Dirac

1. Magnetic Monopoles—Dirac 1931
2. Relativistic "many-time theory"—Dirac 1932; Dirac, Fock, and Podolsky 1933. Generalized by Sin-itiro Tomonaga and Julian Schwinger, as cited by each of them in their Nobel Addresses.
3. The Lagrangian in Quantum Mechanics—Dirac 1933. Generalized by Richard Feynman, as cited by him in his Nobel Address. (However, Dirac was not a believer in renormalization theory.)
4. From 1935 on: Another bridge, namely that between quantum mechanics and general relativity.
5. From 1937 on: Are the fundamental constants changing with cosmological time?

References

Bethe, H.A., Dirac, P.A.M., *et al.*
1989: *from a life of physics*, (Singapore)
Dalitz, R.H. and Peierls, R.
1986: "Paul Adrian Maurice Dirac", *Biographical Memoirs Fellows of the Royal Society* 32, p.139-85.
Darrigol, O.
1990: "Dirac, Paul Arian Maurice", *Dictionary of Scientific Biography, Supp. II*, pp.224-33.
Dirac, P.A.M.
1927a: "The physical interpretation of quantum dynamics", *Proc. Roy. Soc.*A113, pp. 621-41.

1927b: "The quantum theory of emission and absorption of radiation," *Proc. Roy. Soc.*A114, pp. 243-65.
1928: "The quantum theory of the electron, I," *Proc. Roy. Soc.* A128, pp. 610-24.
1930: *The Principles of Quantum Mechanics* (Cambridge, England).
1931: "Quantized singularities in the electromagnetic field," *Proc. Roy. Soc.* A133, pp. 60-72.
1932: "Relativistic quantum mechanics," *Proc. Roy. Soc.*A136, pp. 453-64.
1933: "The Lagrangian in quantum mechanics," *Physikalische Zeitschrift der Sowjetunion* 3, p. 64-72.
1977: "Recollections of an exciting era", in *History of Twentieth Century Physics*, edited by C. Weiner (New York).
1995: *The Collected Works of P.A.M. Dirac, 1942-1948,* edited by R.H. Dalitz (Cambridge)
Dirac, P.A.M., Fock, V.A., and Podolsky,B.
1932: "On quantum electrodynamics", *Physikalische Zeitschrift der Sowjetunion* 2, p. 468-79.
Fermi, E.
1962: *Collected Papers, Volume I* (Chicago).
Kragh, H.S.
1990: *Dirac, a Scientific Biography* (Cambridge)
Kursunoglu, B.N. and Wigner, E.P. (editors:
1987: *Reminiscences about a great physicist: Paul Adrian Maurice Dirac* (Cambridge)
Mehra, J. and Rechenberg, H.
1982: *The Historical Development of Quantum Theory, Volume 4* (New York)
Pais, A., Jacob, M., Olive, I., and Atiyah, M.F.
1998: *Paul Dirac, The Man and his Work* (Cambridge)
Salam, A. and Wigner, E.P.
1972: *Aspects of Quantum Theory* (Cambridge)
Schweber, S.S.
1994: QED *and the Men Who Made It: Dyson, Feynman, Schwinger and Tomonaga* (Princeton).

Leopold Halpern

Proceedings of the Dirac Centennial Symposium
Howard Baer and Alexander Belyaev

From Reminiscences to Outlook

Leopold Halpern
Florida State University
Department of Physics
Tallahassee, FL 32306, USA
E-mail: halpern@phy.fsu.edu

Forty years ago when I had the occasion to meet Dirac for the first time personally at a meeting on gravitational physics in Jablona, Poland, his book on quantum mechanics had already become something of an oasis in my self study of the subject. The unsurpassable clarity and simplicity of his presentation I much later again experienced during the ten years which I spent as his senior research associate at Florida State University in Tallahassee. We regularly had to discuss sophisticated new problems and I rarely needed to ask a further question after his explanation. The proof-reading of his book on relativity revealed to me only one single omission—a missing dot— and he was rather upset when I told him.

I saw Dirac frequently at meetings after Jablona, notably at a meeting on experiments on gravitation, followed by a summer school on the history of physics held at a renaissance palace on the shore of Lake Como in Italy. I remember warning him of the slippery marble stairs leading from the magnificent garden into the lake. I mentioned on this occasion that I had swum across the lake some years before and had planned to swim again this year. The boat traffic on the lake had however swollen with huge freight barges, a hydrofoil and numerous speedboats threatening the swimmer so that I felt rather discouraged from swimming. The following day when many participants had gathered for bathing at noon, Dirac asked me whether I had made the swim. Many other people were interested whether somebody would make the trip of several hours. I decided to swim during the meal hours when nobody would notice and I left my sandals at the water, walking only a few steps barefoot with the crowd toward the dining room in order not to create attention. Nobody took notice of that except Dirac, who turned round and asked: "Halpern, are you going to swim the lake? I nodded and he wished me good luck. After the meal he went to his room and observed me swimming with his binoculars. When I returned, he was at the water and congratulated me.

Dirac was not indifferent toward other peoples' concerns; he actually liked to help them with advice drawn from his own deep insights, but he expressed himself as briefly as possible so that most people paid no attention to it; hence his famous "inequality" according to which there are usually more people who like to speak than people who like to listen. He was rather pragmatically oriented in choosing his habits. He pointed out to me that a lecture in which one is not very interested acts as a fairly strong sleeping pill and he rarely missed an occasion to make use of this harmless sedative, but he never accepted stimulants in food and drink such as alcohol or even strong tea. He expressed the view that the best state of health would provide the best results in one's thinking activity.

In Tallahassee he was one of the few people regularly seen walking between the university and his home on a hill nearby. I often used to drive him in my small beetle car to go on wilderness excursions. Once I had spilled a vast amount of milk on the floor of my car and was worried that the Florida heat would soon transform it into odorous cheese. "Let the dog into the car" was his brief practical advice.

I had soon found out that Tallahassee was quite close to water wilderness areas with spring water so clean that it could be drunk. Wakulla river below the fence on Road 61 was our particular attraction and I came to know every wild animal in this area and frequently organized picnics to a small island with a considerable snake and insect population.

On one occasion I was conveying everyone in my car and then in my canoe. Before the picnic we would swim in the cool water, a moment that Dirac awaited impatiently. On this particular day Prof. G. Marx from Budapest and Cheryl Spencer, an experimental physicist, were with us. I usually went into the water first, followed by Dirac and I used to look back to make sure that he followed me safely through the numerous broken branches until he was able to swim by himself. On this day, however, something very unusual happened; I noticed a brown water snake just between Dirac's legs. Most of the snakes on the island were venomous water snakes. These snakes I had come to know as extremely good natured if not provoked. I turned to Paul and told him that obviously a water moccasin snake was between his legs but there was no reason to panic; he should just walk without any hasty motion back to the island shore, but Dirac adopted his own view about the matter. He wanted first of all to verify whether my statement was correct. Fortunately I found out very soon that it was not; the snake between Dirac's legs was a brown water snake which is not venomous but aggressive and could cause infectious bites. It was sucking in air from the surface of the water. Nevertheless, we still had a beautiful time at the island.

We regularly traversed with the canoe sidearms of the river that were covered by vegetation. Dirac had such a developed sense of orientation that we never needed orientation equipment. He also used to point out the path of least resistance for the canoe against the current. He liked sporting challenges. Once, when aged 76, he met someone on a lake with a motorboat and asked him whether he could try what water-skiing was like. "Paul is still very immature!" his horrified wife then said to me.

Dirac, like me, was a cold weather person. He pointed out to me that he could think much better when it was cool. It was out of consideration for his family that he agreed to move to a hot climate. We had so many points of harmony in science and life that a clever journalist traced this to parallels in our upbringing; my father had also been a teacher in a higher school with preferred subjects not closely related to my scientific interests. I had hardly ever difficulties of communication or exchange of views with Dirac but it took me quite a time until I found out that he had a great affinity to classical music.; he would seldom reveal some of his preferences.

Dirac proved to be a sincere friend once he had opened up to another person. He proved this when Kapiza was prevented by Stalin from returning to Cambridge. Dirac then frequently traveled to Russia to visit him. He developed a liking for the physicists and for the country. He however never agreed to publish a paper together with another theorist (only with experimenters like Schwartz who checked his theoretical results). Dirac told me that his name was added without his consent to the renowned paper with Fock and Podolsky after he had listened to their arguments without making any comments.

I knew that he had given very few people references and I only asked him to write one for me when it could not be avoided. He then said I should write one for myself and give it to him for signature. I thus wrote the reference as short and unassuming as possible but he then made his own modifications, stating also that I stayed at the rather low ranking position that he had available for me from the university because it gave me the possibility to expand my own research in physics. This was true. My position had to be renewed each year and I had not intended to stay too long in a climate that was difficult for me, but Dirac said each time that I was very useful to him and whether I would not like to continue my stay. I knew that he resented being photographed; I had previously filmed wildlife but had lost nearly all my films in a fire. I did not try to film him after this loss but visitors often asked me to film them together with Dirac with their or with my camera. I did so, sometimes standing in front of the canoe in deep water; Prof. G. Marx should have a good film of this kind.

Dirac was very considerate of other people who took care not to interfere with his principles and habits. One of the few differences I had with him concerned the time of our excursions. He used to work every day of the week and make excursions on weekends when, as is well known, he made his discoveries. I had the established habit of using the uninterrupted time of the weekend for my work and making the excursions on weekdays when I would be more disturbed in my office. This was difficult to change for both of us, but Dirac was very considerate and usually agreed to the change.

Dirac had shared the Nobel Prize with Schroedinger; I had been Schroedinger's assistant for three years. I mentioned to Dirac that Schroedinger remarked to me that his views on quantum physics agreed relatively the best with those of Dirac. Dirac thought for a while and then said that he might agree with the statement but the emphasis should be on "relatively." Dirac liked to think back on Schroedinger. I wrote historical studies about the scientists I have known, only when misconceptions about them had been spread (as in the case of Marietta Blau), or when I had been specially urged to do so. This happened in the case of an eulogy right after the death of Dirac[1] and in my lecture entitled "Observations on two of our brightest Stars" which compares my experiences with Schroedinger and Dirac.[2] The latter article, written during my stay at JPL, was after my return to FSU credited to me as a well cited work on astrophysics! Dirac's dry humor and virtuosity in presenting anecdotes deserves an article to itself. He mastered French and German perfectly. He admired the achievements of German physicists and communicated often in German until Hitler came to power; after that he never used German anymore.

He had the greatest admiration for Einstein. When Einstein died there was an unusual moment when he showed his feelings by crying. He liked to point out that Einstein, in his derivation of Planck's radiation formula was actually the first to use matrix elements to calculate transition probabilities, long before Heisenberg. Dirac's health at the time of the Einstein centennial was not very good but he was ready to travel to any Einstein memorial conference which invited him as a speaker. He said to me that it was very important for the world to realize how great a man Einstein had been. I never saw Dirac display pride or arrogance. His famous distinction of question and statement as usually told, struck me in this context, until he gave the right interpretation which makes the situation appear quite differently from that which is persistently told.[2]

Dirac all his life had only one functioning kidney. His parents generally refrained from consulting physicians and believed more in home remedies. Some doctors in Tallahassee found out about the kidney condition when he was eighty one and urged his wife to have the non-functional kidney removed as soon as possible. I tried to oppose such a serious operation at his age but his wife was from a physician's family and was determined to follow blindly the doctors' advice. The operation was scheduled for the early summer. Dirac had presented in spring a lecture in Coral Gables at the Orbis Sciencia but had not yet written it up. I visited Dirac the morning before the operation in his hospital room; he was fully alert and active and discussed physics with me. The day before he still had walked as usual from the hill to the university and back to his house without difficulties. I was scheduled to fly the same afternoon to Sweden where colleagues had organized a conference in my field at which I had some special function. The operation lasted for six hours and I learned that the result was positive, but when I returned from my journey and saw Dirac again I almost could not recognize him. He still survived for about one year in this state. I often tried during this period to clarify some physics questions with him but usually without success. I was approached and urged by Kursunoglu to write up his last Coral Gables lecture. I knew Dirac's critical views about renormalization rather well and wrote the paper as close to his statements as possible. I was still able to present it to him and tried to discuss details receiving however only a few useful hints. During the difficult time of his last months a physicists and in particular his wife, Mark and Sandra Greenfield of Florida A and M. University ,were particularly helpful, taking very good care of Dirac in their home when Dirac's wife had to leave her house to travel.

The reader may from the foregoing nourish the impression that my credentials to become Dirac's senior research associate were mainly of practical nature. I would like to counteract such an impression by mentioning first briefly the research I did to understand the role and magnitude of possible quantum effects of gravitation which are compatible with the general theory of relativity. Schroedinger had during his last days still directed my attention to the question of compatibility of gravitation with quantum theory. He said, he believed a new idea is required to understand the apparent lack of compatibility of the phenomena.[3]

My work examined the upper limit of the probabilities of elementary particle processes caused by or directly involved with gravitational fields. The macroscopic formulation of Einstein's principle of equivalence may create the impression that such transitions are at all excluded by the general theory of relativity. Conclusions from the nonlocal character of quantum theory show however that the processes cannot be excluded although the transition probabilities are in general smaller than crude estimates of lowest order may promise. (We consider here only processes in which the gravitational fields are directly involved and do not contribute just statistically, as in the well known cases of white dwarfs and neutron stars or closely below the Planck scale.) One crucial point to consider is the extreme smallness of the gravitational coupling parameter (the Planck length) to elementary particle energies. An extreme penetrability of high frequency gravitational radiation is thereby implied, and is somewhat further enhanced by the higher spin of gravitons. The gravitational momentum transfer from a heavy body to the domain of an elementary particle process is even for the highest known mass densities experiencing a low cutoff due to the extensions of the gravitating body. Thus for example the decay of one photon into three photons due to such momentum transfers near a large mass remains insignificant compared to related effects due to higher order electromagnetic disturbances.[4]

Gravitational radiation from suitable excited states of elementary particles could for a large number of systems give rise to a substantial energy flux - if there existed not always far more important competing processes from the other interactions, even if of higher order. The electromagnetic radiation from the inside of a hot star is largely confined due to absorption inside the star, whereas the simultaneously produced gravitational thermal radiation of much smaller intensity can escapes due to it's high penetrability. Near the earth orbit we can thus expect a flux of about one graviton of KeV per squared meter and minute from excited Fe atoms inside the sun. The detection of such intensities remains hopeless for gravitons.[5] Collaborating with B. Laurent (the well known scholar of O. Klein) I examined the most varied possibilities of sources of higher frequency gravitational radiation. The absence of promising results led even to hitherto unconsidered possibilities as the construction of a gravitational laser based on quadrupole transitions for which competing electromagnetic avalanches could be prevented by superconducting mirrors transparent to the gravitational radiation. Such a device which we called gazer proved also without chance of a practical realization, chiefly because of the line broadening due to competing transition possibilities. Negative results were also obtained in a work with R. Desbrandes on the emission by superimposed excitations in crystals and by standing electromagnetic waves in a waveguide.[6] There could be no more doubt that the general theory of relativity does not admit any measurable gravitational effects of the kind discussed in the quantum domain. This gives support to Schroedinger's conjecture that this lack is not just a chance, but has unknown deeper reasons. (Schroedinger by the way discovered himself the first quantum effect caused by a gravitational field: the Fermion pair creation in an expanding universe.) Schroedinger's conjecture is rather well in accord with Dirac's ideas which assume two different systems of units of which one, the atomic units are dominated by the quantum of action whereas the other, the gravitational units vary with time relative to them.

I would like to stress that I do not think that the results of my investigation should discourage all experiments on gravitational quantum effects. The results are based on the general theory of relativity of which we know that it does not comprise the results of quantum theory. The first attempts to create a quantum theory of gravitation have now reached already the respectable age of 73 years during which most of the leading theoretical physicists contributed, without that hitherto any striking practical results that one should expect from the universality of the subject ,can be claimed to have been achieved. Some of the pioneers, like Rosenfeld[7] and Schroedinger became very skeptical of a simple minded quantization of the theory of gravitation and also Dirac said that it may remain a problem for a future generation. I had suggested early[5,8] that gravitational radiation could be detected in any kind of matter at low temperature which is shielded from outside electromagnetic sources. The detection demands the observation inside the shielding of two photon dipole transitions stemming from the decay of a quadrupole excitation which must originate from a graviton which has penetrated the shielding from outside. Again, even for rather low frequency gravitons the flux must be enormous to promise a result. We are however hardly able yet to exclude the existence of such huge fluxes by other means. The weakness of the gravitational coupling to quantum systems which we observe at the present epoch allows for cosmological models which could admit such enormous graviton fluxes around us. We know far too little to exclude them by other than gravitational means. Still, the existence of such high weighable graviton fluxes around us is not very likely and I do not recommend to observe the two photon decay in an experiment by itself because of the high expenses. Huge amounts of well shielded matter which is screened for radiation exists however already in the experiments for neutrino detection and it would need only much cheaper supplementation of these experiments for the suggested search of gravitons. The importance of gaining any possible information about the most universal of all interactions cannot be emphasized enough. I have been rather recently approached by two very capable experimenters: R. Desbrandes and D. van Gent who asked me about any possible experiment on gravitational radiation for limited means. The existing theory offers no possibility of this kind and as a consequence no experiment to test it's prediction in the quantum domain has been performed. Any phenomenological hint should if possible be checked by observation - even if it turns out that the field theory in all orders excludes it, as it occurred with a new type of infrared divergence for transitions between photons and gravitons.[9,10]

Dirac has remarked repeatedly that he thinks that the beautiful structure of a simple theory may constitute a more powerful criterion for it's truth than preliminary observational confirmations. When confronted with new astrophysical results that seemed to be in conflict with his Large Numbers Hypothesis, he remained unshaken and reminded me of Einstein's reaction to the experimental results of Kaufmann which seemed to be for a while in disagreement with special relativity. Einstein stated then that he remains convinced of his theory yet he does not want to impose his views on the public. Dirac said he felt quite like this under the given circumstances. I spent two years after Dirac's death with the group of R. Hellings at the Jet Propulsion Laboratory who searched for any confirmations of the prediction of Dirac's Large Numbers Hypothesis. There were no positive results achieved. Analysis of planetary data offered no clue. I tend to believe that Dirac's assumption of the existence of two different systems of units which vary relative to each other with the epoch has much justification. The astrophysical data available by then are continually improved and extended. One will have to reconsider his ideas taking this as far as possible into account.

Dirac's view that the beauty of a theory should be a criterion for it's truth had a profound influence on me. Yet the measure of beauty rests in the eye of the beholder and a developed mind is required to recognize it in a physical theory. I had myself formulated another rule which takes into account that every physical theory is expressed in mathematical language deriving results by logic from axioms, which are idealizations of relations of our experience: it says "Every good (and therefore clear) physical theory will eventually manifest so great an absurdity that no reasonable person can believe in it." I expect this to arise because the objects which the theory describes can only preliminarily be identified with the idealized creations of the axioms; (as for example a planet with a point of space.) The statement should even apply to our best theories as notably the general theory of relativity. I expressed this somewhat more drastically at a Solvay meeting, keeping Schroedinger's discovery of pair creation by gravitational fields and his skepticism toward the physical character of black holes in mind: "There are more things between heaven and the black hole than dreams of in the philosophy of relativist mathematicians." The Hawking radiation was suggested a short time after. There can be no doubt about the observations of massive stars which fulfill all the conditions of general relativity and nuclear physics to predict their collapse to a black hole. Observational details which could provide further crucial information are in general blurred by matter accreting or orbiting around the core. Predictions of detailed further development rest therefore on the (magnificent) construction of Einstein's theory (the physical validity of which in this limit he himself doubted).[18]

I had been thinking since even before my association with Schroedinger about possible broader and deeper generalizations of the gravitational interaction. Schroedinger's formulation of the Dirac electron in gravitational fields had already the formalism (not the intention) of a gauge theory. I hoped to generalize this to include the weak interaction. A breakthrough in my strive to construct a gauge theory which includes the gravitational field happened when I got acquainted with Dirac's formulation of de Sitter and conformal covariant field equations.[11] The extension of the rotation group to the simple Lorentz group had always intrigued me, whereas the further extension to the Poincare" group appeared rather artificial and, later compromised by no go theorems. Dirac's formulation on the manifold of the de Sitter universe (the four-dimensional pseudo-Riemannian space of constant curvature) appealed to me and impressed me by his expression of the field equations solely in terms of generators of the group. I had felt early concern with the precision of the validity of the principle of equivalence; the local limit implied by some formulations of this principle cannot be exact because of the non-locality of quantum phenomena. A spinning test particle in the gravitational fields of general relativity does not move along a geodesic. As most particles have spin, I felt the need for generalizations of the theory which should take this into account. My approach to a generalization of the theory of gravitation is to include the motion of spinning test particles into the principle of inertia by giving it a generalized relativistic form. I make use of the geometry of the simple de Sitter group and postulate: " A particle (structureless or spinning) moves along the projection of an orbit of the de Sitter group on the de Sitter universe unless forces act on it."

The modification of a principle which has been taught in every introduction to physics since the time of Galilei may appear rather sophisticated. I see in it however hardly more than a Carthesian separation into a part that we believe to understand because we constructed it ourself by idealization (Euclidean or de Sitter space) and the remaining part which we don't understand at all and which demands investigation. My modernized version is not so different from the original one except for one feature: the round de Sitter space determines one number which we can express in our units of length and which is extremely large in atomic units- the radius of the universe.

My endeavor to construct a gravitational theory in accord with my version of the principle of inertia led me to work of Bopp and Haag[12] and Neeman and Regge[13] in which the laws of physics were formulated on the whole group manifold. Adoption of this method to my case of the simple de Sitter group proved particularly instructive because of the tight interrelation of all parameters.

The de Sitter universe can be seen as the coset space of the de Sitter group G and it's subgroup L - the Lorentz group: $B = G/L$. A principal fiber bundle: $P(G, L, \Pi; B)$ with $\Pi : G \to B$ the natural projection from points of G onto the base space B. P has all the mathematical structure for a gauge formalism with gauge group L on B. Yang-Mills gauge fields can then only occur if the geometric relations of the group manifold have been "softened". My work differs already in this softening procedure from that of[13] and the school of Torino. Castellani[14] breaks the teleparallelism that exists on the group manifold by softening each component of the operator of the Maurer - Cartan equations into that of a curvature form (or Yang-Mills field). A resulting Poincare gauge theory of the gravitational field has then for topological reasons only derivatives up to the second order of the metric. My work emerges already in 1978 from the Cartan-Killing metric which is defined on the manifold of every simple group. The metric projected from it onto the factor space B is that of the de Sitter universe. I remarked then that the Cartan-Killing metric fulfills Einstein's equations with a cosmological member which equals unity in the present case. (This is the length which I mentioned before). This invites a Kaluza-Klein formalism (somewhat modified by the cosmological member) on the group manifold.

The metric Kaluza-Klein formalism is equivalent to the gauge formalism on the principal fiber bundle touched before. Einstein-Yang-Mills equations result in a similar way as in the work of Klein.[15] The gauge group is however now the six-parameter non - compact Lorentz group which is a subgroup of the general linear group in four dimensions. The equations of the Yang-Mills field are in such a case expressible in terms of equations for a curvature tensor.[16] Depending of how much we soften the group manifold, this tensor maybe but need not be the Riemann tensor. The linear connection may differ from the Christoffel connection by terms with contortion, but it is always a metric connection.

We denote the Riemann-Christoffel tensor by the symbol R_{hijk} and the more general curvature tensor which corresponds to the Yang-Mills fields by the symbol F_{hijk}. The Einstein-Yang-Mills equations expressed by tensors become:

$$F^{h}_{ijk;h} = 0 \tag{1a}$$

$$R_{ik} - \frac{1}{2} g_{ik}(R+1) - (F_{ihjm}F_{k}^{hjm} - \frac{1}{4} g_{ik} F_{abcd} F^{abcd}) = 0 \tag{1b}$$

In the case where $F = R$, the curvature tensor F is identical with the Riemann tensor, eqn. (1a) is equivalent to the equations which C.N. Yang has proposed for a gauge theory of gravitation.[17] All solutions of Einstein's vacuum equations are also solutions of equation (1a) but there are also other solutions of these equations which were considered unphysical and the theory therefore has been discarded. We shall reconsider them together with the set of equation (1b). What I have done to obtain equations (1a,1b) is equivalent to:

1) Consider on the group manifold more general solutions of Einstein's equations with the cosmological member than the Cartan-Killing metric of the group. These solution must have six Killing vector fields with the commutation relations of the Lorentz group.
2) Write the metric linear connection in an orthonormal frame of which six of the components point in direction of the Killing vector fields and the rest is perpendicular to them. Relate these components of the linear connection in this frame to a Yang-Mills field which has the Lorentz group as gauge group.
3) Write the Einstein-Yang-Mills equations on the base (with the appropriate cosmological member).

4) substitute the curvature tensor F for the Yang-Mills field on the manifold B (the de Sitter universe) as outlined earlier.

I have followed the Kaluza-Klein procedure up to step 4). The expression of the Yang-Mills field in terms of the curvature tensor of a metric geometry which is either equal or at least closely related to the Riemannian geometry is of course only posssible if the gauge group is a subgroup of the general linear group. The presence of the cosmological member witnesses the tight interrelations of the simple group (The Lie algebra of a simple group may be compared to a perfect gear box of a top car).

Let us now look at the case where we exclude torsion . Equations (1a,1b) are then fully expressible in terms of the metric and the Riemann tensor. The term bilinear in the curvature tensor of eqn (1b) which corresponds to the energy-momentum tensor of the Yang-Mills field vanishes then for every solution of the geometry which fulfills the Einstein equations with cosmological member. This means that solutions of the vacuum equations of general relativity are somewhat pathologic in this formalism - They have no energy- momentum that would constitute a source of the Einstein term. A possible interpretation would be that these are "phantom solutions" that should be excluded from the physical theory. This fits rather well to my old fashioned and heretic attitude (nourished by my teacher Schroedinger and by Einstein and Rosen[18] which regards singularities in physics mainly due to a lack in our knowledge (manifested by the forementioned weakness of the axioms) and refuses to indulge in the "Grottenbahnromantics" of the collapse to a point. Any decent Yang- Mills field of spin one should somewhere give rise to repulsion. This must have contributed to the rejection of Yang's gauge theory of gravitation which lacks the Einstein term of eqn (1b) which results in attraction, at least as long as the bilinear term in the curvature is small. The simple group chosen results however in the cosmological member and the units of length that we choose for the curvature. These are such that the bilinear term becomes soon important - if it does not vanish. The assumption that purely gravitational fields without the presence of matter have no place in a cosmological theory makes sense. Still then the puzzle remains why the vacuum solutions of general relativity constitute such an excellent approximation to reality within the domain where we live.

The theoretical structure as I have presented it hitherto may serve as a model for a generalized gravitational theory but it does certainly not yet fulfill all the conditions demanded by my version of the principle of inertia because it does not account for the spin precession which must occur together with the deflection of the orbit due to the spin. The only geometric way to remedy this without harm to reality which I have found hitherto, is to introduce contortion terms on the fibers of the bundle which are functions of the curvature on the base. This results in scalar fields on the base (similar as in the Jordan-Brans Dicke theory but which are here functions of the curvature). This way higher nonlinearities in the curvature occur in the equations.

I have tried to outline an attempt to adjust the geometrical gravitational law to the existence of spin. This is to be distinguished from identifying spin with geometrical constructions like torsion as has repeatedly been suggested. Mach's principle is profoundly related to the theoretical structure.

The mathematical structure of the theory has other features which I consider desirable (if not compulsory) for a physical theory. The left and right invariant vectors of the group G imply the existence of inner and outer manifestations of dynamical variables (notably spin and angular momentum) which are usually taken for granted. I can here not go into more details and must refer to past lectures and a future publication.

There is however another point that I must still rise. I had numerous occasions to present my theoretical approach to Dirac and also to Wigner. Dirac had formulated his theory for the de Sitter group as well as for the anti de Sitter group.[11]) Wigner had certain preferences for the anti de Sitter group. This has influenced me to mention my theory only in connection with the anti de Sitter group in some publications. The theory is however constructed to be fitted to each of the two groups. Since the crucial work of the Berkeley group on the determination of cosmic distances from type one supernova candles we have evidence for a cosmological member which is only in accord with the de Sitter group.[19] I consider this as a considerable encouragement for my de Sitter covariant construction and my previous preference for anti de Sitter as (one of) my greatest blunders.

After spending two years at the Jet Propulsion Laboratory I returned to Florida. One of the first things I did here was to revisit the wild river where we had spent so wonderful times with Dirac and his guests. I found all changed. A developer had bought up the wetland for several miles along. The riverfront was parceled into hundred foot stretches. Water cypresses in the river estimated of up to thousand years of age were cut down. Oversize houses built, bulldozers and machine saws roared, dogs barked frightening away the once abundant wildlife. I still found our island but the friendly, noise sensitive snakes had all left; occasionally a less sensitive alligator had remained. Still there were some very big ants that were always friendly. I had let them crawl over me and nibble on any skin impurities. They gently performed true skin operations. I helped every visitor to get used to them and when leaving and even only one was still found in the canoe we used to return and help it to rejoin it's tribe. I had now to find out that even these insects had changed their nature with the conditions and aggressively inflicted me painful bites. The wetland— the soul of the river— had to be filled to allow the planting of lawn. Oversize landing bridges were soon occupied with huge motorboats. Industry was encouraged further downriver and I was shocked by reports of mercury pollution of the hitherto crystal clear drinkable spring water. Many of the people who could afford to buy some of the developed land were nature lovers, who imagined to move to life in the wilderness- yet such development is not long compatible with wildlife and the only creatures that flourish with it are cockroaches and rats and the human analogue of these. Still, there remains a large part of the river in the hands of the state which has retained it's pristine character at the price of admitting visitors only with organized tours. Responsible people are fighting to protect this treasure from the fate the rest have suffered so suddenly. My hope is that also in the future visitors will be able to see some of the characteristics of the unique place that Dirac had enjoyed so much.

Acknowledgements: The plan for the sketched generalization of the gravitational theory originated already in 1978 and besides from Dirac I received also much encouragement from the late William Fairbanks and from Francis Everitt of Stanford University. I am grateful to Maureen Jackson, Billie Oakes and Audrey Wilson for their great help with the language of the text. I thank Prof. R. Jantzen from Mathematical Dept. of Villanova University and Prof. E. Klassen from Math. Dept. of Florida State University for numerous instructive discussions and Mr. Ken Ford from Physics Dept. of Florida State University for his help in preparing my article. I was glad to learn of a different approach to gravitational theory describing the cosmological data during my recent visit to Dubna.[20]

References

1. L. Halpern, Found. Phys. **15**, 257 (1985).
2. L. Halpern, in Differential Geometr. Methods in Theor. Physics, p.463, K. Bleuler and M. Werner Edits. Kluwer Acad.Publ. (1988)
3. L. Halpern, Found. Phys. **11**, 1113 (1987)
4. L. Halpern, Nuov. Cim. **25**, 1239 (1962)
5. L. Halpern and B. Laurent, Nuov. Cim. **33**, 728 (1964)

6. L. Halpern and R. Desbrandes, Ann. Inst. H. Poincare, **11**, 309 (1969); Ondes and Rad. Gravitationelle CNRS Colloq. Nr 220 p. 373-378 (1973)
7. L. Rosenfeld Ann. Phys. (Leipzig) **5**, 311 (1930); Nucl. Phys. **40**, 353 (1963)
8. L. Halpern, Nature, **A88** Letter Nr 128 Phy H7020 Sept. 6, 1971
9. L. Halpern, Ark. f. Fysik, **35** 57 (1967)
10. L. Halpern and B. Jouvet, Ann. Inst. Poincare, A VIII p 25 (1968)
11. P.A.M. Dirac, Annals of Mathem. **36**, 657(1935)
12. F. Bopp and R. Haag, Z. f. Natturforsch. **5a**, 644 (1950)
13. Y. Neeman and T. Regge, Nuov. Cim. **1N5**, 1(1978)
14. L. Castellani, Int. J. Mod. Phys. A **7**, 1583 (1992).
15. O. Klein, Z. Phys. **37**, 895 (1926) [Surveys High Energ. Phys. **5**, 241 (1986)].
16. Y. Choquet, C. De Witt-Morette, M.D. Bleick, "Analysis, Manifolds & Physics" p 378-380 Vbis N. Holland (1982)
17. C.N.Yang, Phys. Rev. Lett. **33**, 445 (1974)
18. A. Einstein, B. Podolsky and N. Rosen, Phys. Rev. **47**, 777 (1935);
19. S. Perlmutter *et al.*, Phys. Rev. Lett. **83**, 670 (1999);
A. Goobar, G. Goldhaber *et al.*, Phys. Scripta **T85**, 47 (2000).
20. D. Behnke, D. B. Blaschke, V. N. Pervushin and D. Proskurin, Phys. Lett. B **530**, 20 (2002).

Monica Dirac

Monica Dirac

Proceedings of the Dirac Centennial Symposium
Howard Baer and Alexander Belyaev

My Father

Monica Dirac

As all of those present who knew my father will remember that he was a very shy man and he always avoided interviews with the press. He refused to talk to reporters and avoided having his photograph taken when ever possible. I think that this is probably why so many published accounts of his life failed to capture his personal side, especially one published in the last decade by the Scientific American. For this reason, I am happy to have the opportunity to share with you some of my memories of him, and to try and put the record straight.

Probably the earliest memory of my father is of him teaching my sister to catch a ball in our back garden. I could not have been more than three years old at the time. Another early memory was of visiting his office at the Arts School in Cambridge, next to the Cavendish Laboratory, where he would go before and after lectures, or to meet his students, and drawing, on what seemed to me at the time, the most enormous black-board.

My father was a very quiet, gentle man who hardly ever got angry. About the only time I ever remember him angry was just after World War II. My uncle had sent us tulip bulbs from Holland. They looked beautiful blooming in the front garden. We had had some cut flowers in the house and after they died I picked a big bunch of the tulips to replace them and happily trotted inside with the bouquet of flowers. I was genuinely surprised at how angry all the grownups were.

My father had many interests apart from physics. He liked to read. Even though he read slowly. He particularly enjoyed science fiction such as H. G. Wells, and Hoyle, who was a contemporary of his at Cambridge. He liked mystery novels such as Sherlock Holmes, Edgar Allan Poe and spy stories such as John Le Carre's.

Another of my childhood memories was rushing downstairs early on Saturday mornings to grab the comics, Beano and Dandy, before my father or sister. We all three were after them and enjoyed reading them. My father told us how his father would confiscate comics from the students at the school where he taught and bring them home for his children to read.

When he was visiting the Institute for Advanced Study in Princeton in the early 1930s, my uncle, Eugene Wigner introduced him to my mother. It was love at first sight and in no time they were courting. My mother introduced him to classical music. I remember as a child walking into the drawing room of our house in Cambridge, in the evening, thinking no one was there as the lights were out, only to find my father sitting quietly in the dark, listening to classical music on the radio. He never went to concerts because he found the audience coughing too distracting. When we bought a television in 1952, in time to watch the coronation, my parents had the shop bring out several different models and my father selected the one with the best sound. If a concert was broadcast simultaneously on the radio and the TV, he would watch it on the TV with the sound turned off and listen to it on the radio.

I remember our family Christmas holidays in England. After Christmas we would stay a week or two in a hotel in Earls Court, in London. My father would take my sister and me to the museums, our favorite was the Science Museum where we could push buttons and turn handles to see exhibits move. In the evenings we would go to see plays or the ballet. My mother adored to watch ballet.

My father came back from a sabbatical in India in the Mid 50s with a print of a Salvador Dali painting. He was intrigued by the Indian artist's work, so was I.

Another of my father's hobbies was playing chess. He enjoyed working through the chess problems in the newspaper.

Everyone who knew my father will remember that he loved to hike. I have been told that on a visit to Russia before World War II, he climbed Mount Elbrus, the highest mountain in the Caucuses, without oxygen and passed out near the top. Every Sunday morning from as long ago as I can remember, my father took my sister and me for a hike or bicycle ride while my mother stayed at home to cook the dinner. He also loved to swim, but he would NEVER swim in a swimming pool, only in rivers, lakes, abandoned quarries, or the sea. He never minded how cold the water was and would swim for so long that when he emerged, he would sometimes be shivering for what seemed to me to be hours! We used to have family outings to the Ooze in Huntingdonshire. We would rent a row boat, row up the river to the millpond, where we would find a spot on the shore to have a pick-nick and then go for a swim in. After my parents moved to Florida, my father enjoyed canoeing on the Wakulla River, and swimming in the sink holes.

Another of my father's hobbies was gardening. Every Saturday and Sunday afternoon when it was not raining he would work in the garden, all 2/3rds of an acre he would maintain himself. My mother would hire gardeners to help but my father would always fire them. He would mow the extensive lawns, tend the flower beds, prune trees, and grow vegetables. At first he mowed the lawn with an old fashioned push mower. Later he graduated to an electric mower with yards and yards of electrical cable. I was always impressed that he NEVER once mowed that cable by mistake. He also harvested fruit from all our many fruit trees. He would pick basket after basket of apples and carefully place them on shelves, on the first floor (English first floor) of the garage, making sure that none were touching. Both my parents were very frugal. My father hated to waste the windfall apples. My mother usually refused to deal with them. So my father would collect them, boil them up, turn the kitchen stool upside down to hang the apple pulp in muslin so the juice could drip into a bowl to make apple jelly. My father always said that the greener the apples the redder the jelly would be. During World War II, I have been told that he grew mushrooms in the cellar, in the garage, and behind the garage where we later had our wood pile. We also had two large asparagus beds. I remember him preparing the second one. He dug a large trench over 3 feet deep, about 4 feet wide and over 20 feet long to remove the clay subsoil. The trench was slowly filled with organic matter, and compost from the garden. He then grew asparagus plants from seed for the new bed. The project took several years to complete. I also remember him growing peas. He would coat the pea seeds with dripping and then role them in red lead oxide powder, to discourage birds from eating the newly emerged pea seedlings. People were less conscious of environmental health hazards in those days.

I have always loved animals and never ceased trying to persuade my parents to let me have more pets. My father did not like dogs, he did not like being startled when they barked. Cats were better. I remember one black cat we had. The cat used to go in and out of the house through the shoot where coal was delivered to be stored in a small room in the cellar, next to the furnace. My father wanted to board up the hole but still leave a hole large enough for the cat. So he asked me to bring him the cat, and he measured the distance between the tips of its whiskers to ensure that he was leaving sufficient space.

My father was always willing to help me with my maths or science homework. He would explain the problem in such generic terms that I never had the problems that the other children had, of the teacher complaining that they had used the 'wrong method' to solve the problem. My father always said that when outlining a new concept in a lecture, one should always repeat oneself several times, preferably using different words.

When I became interested in collecting fossils and mineral specimens as a teenager, he was always willing to go with me and to support my hobbies. But when I asked for advice about important matters such as 'what should I be when I grew up' he would never give advice and left me to make up my own mind. This was probably because his elder brother had wanted to be a doctor, but his father insisted that he become an electrical engineer. That plus trouble with his girl-friend lead his brother to commit suicide in his early twenties.

My father was a strict tee-totaller. If my sister or I had an upset stomach, my mother liked to give us a small glass of cognac but my father would be furious if he found out. He never liked to eat any food cooked with wine or sherry. When I made the hard sauce to go with our Christmas pudding it was always a challenge, how much alcohol can I put in without arousing my father's suspicions? My father never had tea or coffee until he was 21. He liked his tea very weak. I remember an occasion when we were staying in the Schrödinger's house in Dublin and had been invited to tea by the Guinness family. My father asked for weak tea without milk and sent it back to our hostess four times because it was too strong! He was also suspicious of pickles. He would rarely allow my sister or me to eat pickles with our cold meat, he thought they were bad for children and would cause ulcers.

I would like to end by repeating that although my father was quiet and shy, he had many interests outside his work. He enjoyed spending time with his family, he loved to travel, and every summer we took long family vacations. I clearly remember him saying that no one can work hard on a serious intellectual problem for more than 4 hours a day.

Frank Wilczek

Proceedings of the Dirac Centennial Symposium
Howard Baer and Alexander Belyaev

The Dirac Equation

Frank Wilczek
Center for Theoretical Physics
Massachusetts Institute of Technology
Cambridge, MA 02139-4307

> One cannot escape the feeling that these mathematical formulae have an independent existence and an intelligence of their own, that they are wiser than we are, wiser even than their discoverers, that we get more out of them than was originally put into them.
>
> — *H. Hertz, on Maxwell's equations for electromagnetism*

> A great deal of my work is just playing with equations and seeing what they give.
>
> —*P.A.M. Dirac*

> It gave just the properties one needed for an electron. That was really an unexpected bonus for me, completely unexpected.
>
> —*P.A.M. Dirac, on the Dirac equation*

Of all the equations of physics, perhaps the most "magical" is the Dirac equation. It is the most freely invented, the least conditioned by experiment, the one with the strangest and most startling consequences.

In early 1928 (the receipt date on the original paper is January 2), Paul Adrien Maurice Dirac (1902–1984), a 25-year-old recent convert from electrical engineering to theoretical physics, produced a remarkable equation, forever to be known as the Dirac equation. Dirac's goal was quite concrete, and quite topical. He wanted to produce an equation that would describe the behavior of electrons more accurately than previous equations. Those equations incorporated either special relativity or quantum mechanics, but not both. Several other more prominent and experienced physicists were working on the same problem.

Unlike these other physicists, and unlike the great classics of physics, Newton and Maxwell, Dirac did not proceed from a minute study of experimental facts. Instead he guided his search using a few basic facts and perceived theoretical imperatives, some of which we now know to be wrong. Dirac sought to embody these principles in an economical, mathematically consistent scheme. By "playing with equations," as he put it, he hit upon a uniquely simple, elegant solution. This is, of course, the equation we now call the Dirac equation.

Some consequences of Dirac's equation could be compared with existing experimental observations. They worked quite well, and explained results that were otherwise quite mysterious. Specifically, as I'll describe below, Dirac's equation successfully predicts that electrons are always spinning and that they act as little bar magnets, and the rate of the spin and the strength of the magnetism. But other consequences appeared utterly inconsistent with obvious facts. Notably, Dirac's equation contains solutions that appear to describe a way for ordinary atoms to wink out into bursts of light, spontaneously, in a fraction of a second.

For several years Dirac and other physicists struggled with an extraordinary paradox. How can an equation be "obviously right" since it accounts accurately for many precise experimental results, and achingly beautiful to boot – and yet manifestly, catastrophically wrong?

The Dirac equation became the fulcrum on which fundamental physics pivoted. While keeping faith in its mathematical form, physicists were forced to reexamine the meaning of the symbols it contains. It was in this confused, intellectually painful re-examination – during which Werner Heisenberg wrote to his friend Wolfgang Pauli, "The saddest chapter of modern physics is and remains the Dirac theory" and "In order not to be irritated with Dirac I have decided to do something else for a change⋯" – that truly modern physics began.

A spectacular result was the prediction of *antimatter* – more precisely, that there should be a new particle with the same mass as the electron, but the opposite electric charge, and capable of annihilating an electron into pure energy. Particles of just this type were promptly identified, through painstaking scrutiny of cosmic ray tracks, by Carl Anderson in 1932.

The more profound, encompassing result was a complete reworking of the foundations of our description of matter. In this new physics, particles are mere ephemera. They are freely created and destroyed; indeed, their fleeting existence and exchange is the source of all interactions. The truly fundamental objects are universal, transformative ethers: quantum fields. These are the concepts that underlie our modern, wonderfully successful Theory of Matter (usually called, quite inadequately, the Standard Model). And the Dirac equation itself, drastically reinterpreted and vastly generalized, but never abandoned, remains a central pillar in our understanding of Nature.

7. Dirac's Problem and the Unity of Nature

The immediate occasion for Dirac's discovery, and the way he himself thought about it, was the need to reconcile two successful, advanced theories of physics that had gotten slightly out of synch. By 1928 Einstein's special theory of relativity was already over two decades old, well digested, and fully established. (The general theory, which describes gravitation, is not part of our story here. Gravity is negligibly weak on atomic scales.) On the other hand, the new quantum mechanics of Heisenberg and Schrödinger, although quite a young theory, had already provided brilliant insight into the structure of atoms, and successfully explained a host of previously mysterious phenomena. Clearly, it captured essential features of the dynamics of electrons in atoms. The difficulty was that the equations developed by Heisenberg and Schrödinger did not take off from Einstein's relativistic mechanics, but from the old mechanics of Newton. Newtonian mechanics can be an excellent approximation for systems in which all velocities are much smaller than the speed of light, and this includes many cases of interest in atomic physics and chemistry. But the experimental data on atomic spectra, which one could address with the new quantum theory, was so accurate that small deviations from the Heisenberg-Schrödinger predictions could be observed. So there was a strong "practical" motivation to search for a more accurate electron equation, based on relativistic mechanics. Not only young Dirac, but also several other major physicists, were after such an equation.

In hindsight we can discern that much more ancient and fundamental dichotomies were in play: light versus matter; continuous versus discrete. These dichotomies present tremendous barriers to the goal of achieving a unified description of Nature. Of the theories Dirac and his contemporaries sought to reconcile, relativity was the child of light and the continuum, and quantum theory the child of matter and the discrete. After Dirac's revolution had run its course, all were reconciled, in the mind-stretching conceptual amalgam we call a quantum field.

The dichotomies light/matter and continuous/discrete go deep. They were experienced by the earliest sentient proto-humans. They were articulated clearly, and debated inconclusively, by the ancient Greeks. Specifically, Aristotle distinguished Fire and Earth as primary elements - light versus matter. And he argued, against the Atomists, in favor of a fundamental plenum ("Nature abhors a vacuum") - upholding the continuous, against the discrete.

These dichotomies were not relieved by the triumphs of classical physics; indeed, they were sharpened.

Newton's mechanics is best adapted to describing the motion of rigid bodies through empty space. While Newton himself in various places speculated on the possible primacy of either side of both dichotomies, Newton's followers emphasized his "hard, massy, impenetrable" atoms as the fundamental building-blocks of Nature. Even light was modeled in terms of particles.

Early in the nineteenth century a very different picture of light, according to which it consists of waves, scored brilliant successes. Physicists accepted that there must be a continuous, space-filling ether to support these waves. The discoveries of Faraday and Maxwell, assimilating light to the play of electric and magnetic fields, which are themselves continuous entities filling all space, refined and reinforced this idea.

Yet Maxwell himself, and Ludwig Boltzmann, succeeded in showing that the observed properties of gases, including many surprising details, could be explained if the gases were composed of many small, discrete, well-separated atoms moving through otherwise empty space. Furthermore J.J. Thomson experimentally, and Hendrik Lorentz theoretically, established the existence of electrons as building-blocks of matter. Electrons appear to be indestructible particles, of the sort that Newton would have appreciated.

Thus as the twentieth century opened, physics featured two quite different sorts of theories, living together in uneasy peace. Maxwell's electrodynamics is a continuum theory of electric and magnetic fields, and of light, that makes no mention of mass. Newton's mechanics is a theory of discrete particles, whose *only* mandatory properties are mass and electric charge[a].

Early quantum theory developed along two main branches, following the fork of our dichotomies, but with hints of convergence.

[a]That is, to predict the motion of a particle you need to know its charge and its mass: no more, no less. The value of the charge can be zero; then the particle will have only gravitational interactions.

One branch, beginning with Planck's work on radiation theory, and reaching a climax in Einstein's theory of photons, dealt with light. Its central result is that light comes in indivisible minimal units, photons, with energy and momentum proportional to the frequency of the light. This, of course, established a particle-like aspect of light.

The second branch, beginning with Bohr's atomic theory and reaching a climax in Schrödinger's wave equation, dealt with electrons. It established that the stable configurations of electrons around atomic nuclei were associated with regular patterns of wave vibrations. This established a wave-like property of matter.

Thus the fundamental dichotomies softened. Light is a bit like particles, and electrons are a bit like waves. But sharp contrasts remained. Two differences, in particular, appeared to distinguish light from matter sharply.

First, if light is to be made of particles, then they must be very peculiar particles, with internal structure, for light can be polarized. To do justice to this property of light, its particles must have some corresponding property. There can't be an adequate description of a light beam specifying only that it is composed of so-and-so many photons with such-and-such energies; those facts will tell us how bright the beam is, and what colors it contains, but not how it is polarized. To get a complete description, one must also be able to say which way the beam is polarized, and this means that its photons must somehow carry around arrows that allow them to keep a record of the light's polarity. This would seem to take us away from the traditional ideal of elementary particles. If there's an arrow, what's *it* made of? – and why can't it be separated from the particle?

Second, and more profound, photons are evanescent. Light can be radiated, as when you turn on a flashlight, or absorbed, as when you cover it with your hand. Therefore particles of light can be created or destroyed. This basic, familiar property of light and photons takes us far away from the traditional ideal of elementary particles. The stability of matter would seem to require indestructible building-blocks, with properties fundamentally different from evanescent photons.

The Dirac equation, and the crisis it provoked, forced physicists, finally, to transcend all these dichotomies. The consequence is a unified concept of substance, that is surely one of mankind's greatest intellectual achievements.

8. The Early Payoff: Spin

Dirac was working to reconcile the quantum mechanics of electrons with special relativity. He thought – mistakenly, we now know – that quantum theory required equations of a particularly simple kind, the kind mathematicians call first-order. Never mind why he thought so, or precisely what first-order means; the point is that he wanted an equation that is, in a certain very precise sense, of the simplest possible kind. Tension arises because it is not easy to find an equation that is both simple in this sense and also consistent with the requirements of special relativity. To construct such an equation, Dirac had to expand the terms of the discussion. He found he could not get by with a single first-order equation – he needed a system of four intricately related ones, and it is actually this system we refer to as "the" Dirac equation.

Two equations were quite welcome. Four, initially, were a big problem.

First, the good news.

Although the Bohr theory gave a good rough account of atomic spectra, there were many discrepant details. Some of the discrepancies concerned the number of electrons that could occupy each orbit, others involved the response of atoms to magnetic fields, as manifested in the movement of their spectral lines. Wolfgang Pauli had shown, through detailed analysis of the experimental evidence, that Bohr's model could only work, even roughly, for complex atoms if there were a tight restriction on how many electrons could occupy any given orbit. This is the origin of the famous Pauli exclusion principle. Today we learn this principle in the form "only one electron can occupy a given state". But Pauli's original proposal was not so neat; it came with some disturbing fine print. For the number of electrons that could occupy a given Bohr orbital was not one, but two. Pauli spoke obscurely of a "classically non-describable duplexity", but – needless to say – did not describe any reason for it.

In 1925 two Dutch graduate students, Samuel Goudsmit and George Uhlenbeck, devised a possible explanation of the magnetic response problems. If electrons were actually tiny magnets, they showed, the discrepancies would disappear. Their model's success required that all electrons must have the same magnetic strength, which they could calculate. They went on to propose a mechanism for the electron's magnetism. Electrons, of course, are electrically charged particles. Electric charge in circular motion generates magnetic fields. Thus, if for some reason electrons were always rotating about their own axis, their magnetism might be explained. This intrinsic *spin* of electrons would have an additional virtue. If the rate of spin were the minimum allowed by quantum mechanics[b], then Pauli's "duplexity" would be explained. For the spin would have no possibility to vary in magnitude, but only the possibility to point either up or down. Many eminent physicists were quite skeptical of Goudsmit and Uhlenbeck. Pauli himself tried to dissuade them from publishing their work. For one thing, their model seemed to require the electron to rotate at an extraordinarily rapid rate, at its surface probably faster than the speed of light. For another, they gave no account of what holds an electron together. If it is an extended distribution of electric charge, all of the same sign, it will want to fly apart – and rotation, by introducing centrifugal forces, only makes the problem worse. Finally, there was a quantitative mismatch between their requirements for the strength of the electron's magnetism and the amount of its spin. The ratio of these two quantities is governed by a factor called the gyromagnetic ratio, written g. Classical mechanics predicts $g = 1$, whereas to fit the data Goudsmit and Uhlenbeck postulated $g = 2$. But despite these quite reasonable objections, their model stubbornly continued to agree with experimental results!

Enter Dirac. His system of equations allowed a class of solutions, for small velocities, in which only two of the four functions appearing in his equations are appreciable. This was duplexity, but with a difference. Here it fell out automatically as a consequence of implementing general principles, and most definitely did not have to be introduced *ad hoc*. Better yet, using his equation Dirac could calculate the magnetism of electrons, also without further assumptions. He got $g = 2$. Dirac's great paper of 1928 wastes no words. Upon demonstrating this result, he says simply

> The magnetic moment is just that assumed in the spinning electron model.

And a few pages later, after working out the consequences, he concludes laconically

> The present theory will thus, in the first approximation, lead to the same energy levels as those obtained by [C.G.] Darwin, which are in agreement with experiment.

His results spoke loudly for themselves, with no need for amplification. From

[b]In quantum mechanics, only certain values of the discrete spin are allowed. This is closely related to the restriction on allowed Bohr orbitals.

then on, there was no escaping Dirac's equation. Whatever difficulties arose – and there were some big and obvious ones – they would be occasions for struggle, not desertion. Such gleaming jewels of insight would be defended at all costs.

Although his intellectual starting point, as I mentioned, was quite different and more abstract, Dirac begins his paper by referring to Goudsmit, Uhlenbeck, and the experimental success of their model. Only in the second paragraph does he reveal his hand. What he says is quite pertinent to the themes I emphasized above.

> The question remains as to why Nature should have chosen this particular model for the electron instead of being satisfied with a point-charge. One would like to find some incompleteness in the previous methods of applying quantum mechanics to the point-charge such that, when removed, the whole of the duplexity phenomena follow without arbitrary assumptions.

Thus Dirac is not offering a new model of electrons, as such. Rather, he is defining a new *irreducible* property of matter, inherent in the nature of things, specifically in the consistent implementation of relativity and quantum theory, that arises even in the simplest possible case of structureless point particles. Electrons happen to be embodiments of this simplest possible form of matter. The valuable properties of Goudsmit and Uhlenbeck's "spin", specifically its fixed magnitude and its magnetic action, which aid in the description of observed realities, were retained, now based on a much deeper foundation. The arbitrary and unsatisfactory features of their model are bypassed.

We were looking for an arrow that would be a necessary and inseparable part of elementary bits of matter, like polarization for photons. Well, there it is!

The spin of the electron has many practical consequences. It is responsible for the phenomenon of ferromagnetism, and the enhancement of magnetic fields in the core of electric coils, which forms the heart of modern power technology (motors and dynamos). Active manipulation of electron spins allows us to store and retrieve a great deal of information in a very small volume (magnetic tape, disk drives). Even the much smaller and more inaccessible spin of atomic nuclei plays a big role in modern technology. Manipulating such spins with radio and magnetic fields, and sensing their response, is the basis of the magnetic resonance imaging (MRI) so useful in medicine. This application, among many others, would be inconceivable (literally!) without the exquisite control of matter that only fundamental understanding can bring.

Spin in general, and Dirac's prediction for the magnetic moment in particular, has also played a seminal role in the subsequent development of fundamental physics. Small deviations from Dirac's $g = 2$ were discovered by Polykarp Kusch and collaborators in the 1940s. They provided some of the first quantitative evidence for the effects of virtual particles, a deep and characteristic property of quantum field theory. Very large deviations from $g = 2$ were observed for protons and neutrons in the 1930s. This was an early indication that protons and neutrons are not fundamental particles in the same sense that electrons are. But I'm getting ahead of the story.···

9. The Dramatic Surprise: Antimatter

Now for the 'bad' news.

Dirac's equation consists of four components. That is, it contains four separate wave functions to describe electrons. Two components have an attractive and immediately successful interpretation, as we just discussed, describing the two possible directions of an electron's spin. The extra doubling, by contrast, appeared at first to be quite problematic.

In fact, the extra equations contain solutions with *negative* energy (and either direction of spin). In classical (non-quantum) physics the existence of extra solutions would be embarrassing, but not necessarily catastrophic. For in classical physics, you can simply choose not to use these solutions. Of course that begs the question why *Nature* chooses not to use them, but it is a logically consistent procedure. In quantum mechanics, even this option is not available. In quantum physics, generally "that which is not forbidden is mandatory". In the specific case at hand, we can be quite specific and precise about this. All solutions of the electron's wave equation represent possible behaviors of the electron, that will arise in the right circumstances. Assuming Dirac's equation, if you start with an electron in one of the positive-energy solutions, you can calculate the rate for it to emit a photon and transition into one of the negative-energy solutions. Energy must be conserved overall, but that is not a problem here – it just means that the energy of the emitted photon would be *more* than that of the electron which emitted it! Anyway, the rate turns out to be ridiculously fast, a small fraction of a second. So you can't ignore the negative-energy solutions for long. And since an electron has never been observed to do something so peculiar as radiating more energy than it starts with, there was, on the face of it, a terrible problem with the quantum mechanics of Dirac's equation.

Dirac was well aware of this problem. In his original paper, he simply acknowledged

> For this second class of solutions W [the energy] has a negative value. One gets over the difficulty on the classical theory by arbitrarily excluding those solutions that have a negative W. One cannot do this on the quantum theory, since in general a perturbation will cause transitions from states with W positive to states with W negative.··· The resulting theory is therefore still only an approximation, but it appears to be good enough to account for all the duplexity phenomena without arbitrary assumptions.

and left it at that. This was the situation that provoked Heisenberg's outbursts to Pauli, quoted earlier.

By the end of 1929 – not quite two years later – Dirac made a proposal to address the problem. It exploited the Pauli exclusion principle, according to which no two electrons obey the same solution of the wave equation. What Dirac proposed was a radically new conception of empty space. He proposed that what we consider 'empty' space is in reality chock-a-block with negative-energy electrons. In fact, according to Dirac, *'empty' space actually contains electrons obeying all the negative energy solutions.* The great virtue of this proposal is that it explains away the troublesome transitions from positive to negative solutions. A positive-energy electron can't go to a negative-energy solution, because there's always another electron already there, and the Pauli exclusion principle won't allow a second one to join it.

It sounds outrageous, on first hearing, to be told that what we perceive as empty space is actually quite full of stuff. But, on reflection, why not? We have been sculpted by evolution to perceive aspects of the world that are somehow useful for our survival and reproductive success. Since unchanging aspects of the world, upon which we can have little influence, are not useful in this way, it should not seem terribly peculiar that they would escape our untutored perception. In any case, we have no warrant to expect that naive intuitions about what is weird or unlikely provide reliable guidance for constructing models of fundamental structure in the microworld, because these intuitions derive from an entirely different realm of phenomena. We must take it as it comes. The validity of a model must be judged according to the fruitfulness and accuracy of its consequences.

So Dirac was quite fearless about outraging common sense. He focused, quite properly, on the observable consequences of his proposal.

Since we are considering the idea that the ordinary state of "empty" space is far from empty, it is helpful to have a different, more non-committal word for it. The one physicists like to use is "vacuum".

In Dirac's proposal, the vacuum is full of negative-energy electrons. This makes the vacuum a medium, with dynamical properties of its own. For example, photons can interact with the vacuum. One thing that can happen is that if you shine light on the vacuum, providing photons with enough energy, then a negative-energy electron can absorb one of these photons, and go into a positive-energy solution. The positive-energy solution would be observed as an ordinary electron, of course. But in the final state there is also a *hole* in the vacuum, because the solution originally occupied by the negative-energy electron is no longer occupied.

The idea of holes was, in the context of a dynamical vacuum, startlingly original, but it was not quite unprecedented. Dirac drew on an analogy with the theory of heavy atoms, which contain many electrons. Within such atoms, some of the electrons correspond to solutions of the wave equation that reside nearby the highly charged nucleus, and are very tightly bound. It takes a lot of energy to break such electrons free, and so under normal conditions they present an unchanging aspect of the atom. But if one of these electrons absorbs a high-energy photon (an X-ray) and is ejected from the atom, the change in the normal aspect of the atom is marked by its *absence*. The absence of an electron, which would have supplied negative charge, by contrast looks like a positive charge. The positive effective charge follows the orbit of the missing electron, so it has the properties of a positively charged particle.

Based on this analogy and other hand-waving arguments – the paper is quite short, and practically devoid of equations – Dirac proposed that holes in the vacuum are positively charged particles. The process where a photon excites a negative-energy electron in the vacuum to a positive energy is then interpreted as the photon creating an electron and a positively charged particle (the hole). Conversely, if there is a preexisting hole, then a positive-energy electron can emit a photon and occupy the vacant negative-energy solution. This is interpreted as the annihilation of an electron and a hole into pure energy. I referred to a photon being emitted, but this is only one possibility. Several photons might be emitted, or any other form of radiation that carries away the liberated energy.

Dirac's first hole theory paper was entitled "A Theory of Electrons and Protons". At the time protons were the only known positively charged particles. It was therefore natural to try to identify the hypothetical holes as protons. But severe difficulties with this identification were soon evident. Specifically, the two sorts of process we just discussed – production of electron-proton pairs, and annihilation of electron-proton pairs – have never been observed. The second is especially problematic, because it predicts that hydrogen atoms spontaneously self-destruct in microseconds – which, thankfully, they do not.

There was also a logical difficulty with the identification of holes with protons. Based on the symmetry of the equations, one could demonstrate that the holes must have the same mass as the electrons. But a proton has, of course, a much larger mass than an electron.

In 1931 Dirac withdrew his earlier identification of holes with protons, and accepted the logical outcome of his own equation and the dynamical vacuum it required:

> A hole, if there was one, would be a new kind of elementary particle, unknown to experimental physics, having the same mass and opposite charge of the electron.

On August 2, 1932, Carl Anderson, an American experimentalist studying photographs of the tracks left by cosmic rays in a cloud chamber, noticed some tracks that lost energy as expected for electrons, but were bent in the opposite direction by the magnetic field. He interpreted this as indicating the existence of a new particle, now known as the antielectron or positron, with the same mass as the electron but the opposite electric charge. Ironically, Anderson was completely unaware of Dirac's prediction.

Thousands of miles away from his rooms at Saint John's, Dirac's holes – the product of his theoretical vision and revision – had been found, descending from the skies of Pasadena. So in the long run the "bad" news turned out to be "even better" news. Negative-energy frogs became positronic princes.

Today positrons are no longer a marvel, but a tool. A notable use is to take pictures of the brain in action – PET scans, for positron-electron tomography. How do positrons get into your head? They are snuck in by injecting molecules containing atoms whose nuclei are radioactive, and decay with positrons as one of their decay products. These positrons do not go very far before they annihilate against some nearby electron, usually producing two photons, which escape your skull, and can be detected. Then you can reconstruct where the original molecule went, to map out metabolism, and you can also study the energy loss of the photons on the way out, to get a density profile, and ultimately an image, of the brain tissue.

Another notable application is to fundamental physics. You can accelerate positrons to high energy, as you can of course electrons, and bring the beams together. Then the positrons and electrons will annihilate, producing a highly concentrated form of "pure energy". Much of the progress in fundamental physics over the past half century has been based on studies of this type, at a series of great accelerators all over the world, the latest and greatest being the LEP (large electron-positron) collider at CERN, outside Geneva. I'll be discussing a stunning highlight of this physics a little later.

The physical ideas of Dirac's hole theory, which as I mentioned had some of its roots in the earlier study of heavy atoms, fed back in a big way into solid state physics. In solids one has a reference or ground configuration of electrons, with the lowest possible energy, in which electrons occupy all the available states up to a certain level. This ground configuration is the analogue of the vacuum in hole theory. There are also configurations of higher energy, wherein some of the low-energy states are not used by any electron. In these configurations there are vacancies or "holes" – that's what they're called, technically – where an electron would ordinarily be. Such holes behave in many respects like positively charged particles. Solid-state diodes and transistors are based on clever manipulation of holes and electron densities at junctions between different materials. One also has the beautiful possibility to direct electrons and holes to a place where they can combine (annihilate). This allows you to design a source of photons that you can control quite precisely, and leads to such mainstays of modern technology as LEDs (light-emitting diodes) and solid-state lasers.

In the years since 1932 many additional examples of anti-particles have been observed. In fact, for every particle that has ever been discovered, a corresponding anti-particle has also been found. There are antineutrons, antiprotons, antimuons (the muon itself is a particle very similar to the electron, but heavier), antiquarks of various sorts, even antineutrinos, and anti-π mesons, anti-K mesons,[c]. Many of these particles do not obey the Dirac equation, and some of them do not even obey the Pauli exclusion principle. So the physical reason for the existence of antimatter must be very general – much more general than the arguments that first led Dirac to predict the existence of positrons.

[c]An interesting case is the photon, which is its own antiparticle. This is not possible for a charged particle, but the photon is electrically neutral.

In fact, there is a very general argument that if you implement both quantum mechanics and special relativity, every particle must have a corresponding antiparticle. A proper presentation of the argument requires either a sophisticated mathematical background or a lot of patience. Here I'll be content with a rough version, which shows why antimatter is a plausible consequence of implementing both relativity and quantum mechanics, but doesn't quite nail the case.

Consider a particle, let's say a shmoo, to give it a name (while emphasizing that it could be *anything*), moving east at very nearly the speed of light. According to quantum mechanics, there is actually some uncertainty in its position. So there's some probability, if you measure it, that you will find that the shmoo is slightly west of its expected mean position at an initial time, and slightly east of its expected mean position at a later time. So it has traveled further than you might have expected during this interval – which means it was traveling more quickly. But since the expected velocity was essentially the speed of light, the faster speeds required to accommodate uncertainty threaten to violate special relativity, which requires that particles cannot move faster than the speed of light. It's a paradox.

With antiparticles, you can escape the paradox. It requires orchestrating a symphony of weird ideas, but it's the only way people have figured out how to do it, and it seems to be Nature's way. The central idea is that, yes, uncertainty does mean that you can find a shmoo where special relativity tells you your shmoo can't be - but the shmoo you observe is not necessarily the same as the one you were looking for! For it's also possible that at the later time there are two shmoos, the original one and a new one. To make this consistent there must also be an anti-shmoo, to balance the charge, and to cancel out any other conserved quantities that might be associated with the additional shmoo. What about the energy balance - aren't we getting out more than we put in? Here, as often in quantum theory, to avoid contradictions you must be specific and concrete in thinking about what it means to measure something. One way to measure the shmoo's position would be to shine light on it. But to measure the position of a fast-moving shmoo accurately we have to use high-energy photons, and there's also then the possibility such a photon will create a shmoo-anti-shmoo pair. And in that case – closing the circle – when you report the result of your position measurement, you might be talking about the wrong shmoo!

10. The Deepest Meanings: Quantum Field Theory

Dirac's hole theory is brilliantly clever, but Nature goes deeper. Although hole theory is internally consistent, and can cover a wide range of applications, there are several important considerations that force us to go beyond it.

First, there are particles that do not have spin, and do not obey the Dirac equation, and yet have antiparticles. This is no accident: the existence of antiparticles is a general consequence of combining quantum mechanics and special relativity, as I just discussed. Specifically, for example, positively charged π^+ mesons (discovered in 1947) or W^+ bosons (discovered in 1983) are quite important players in elementary particle physics, and they do have antiparticles π^- and W^-. But we can't use Dirac's hole theory to make sense of these antiparticles, because π^+ and W^+ particles don't obey the Pauli exclusion principle. So there is no possibility of interpreting their antiparticles as holes in a filled sea of negative-energy solutions. If there are negative-energy solutions, whatever equation they satisfy[d], occupying them with one particle will not prevent another particle from entering the same state. Thus catastrophic transitions into negative-energy states, which Dirac's hole theory prevents for electrons, must be banished in a different way.

Second, there are processes in which the number of electrons minus the number of positrons changes. An example is the decay of a neutron into a proton, an electron, and an antineutrino. In hole theory the excitation of a negative-energy electron into a positive-energy state is interpreted as creation of a positron-electron pair, and de-excitation of a positive-energy electron into an unoccupied negative-energy state is interpreted as annihilation of an electron-positron pair. In neither case does the difference between the number of electrons and the number of positrons change. Hole theory cannot accommodate changes in this difference. So there are definitely important processes in Nature, even ones specifically involving electrons, that do not fit easily into Dirac's hole theory.

The third and final reason harks back to our initial discussion. We were looking to break down the great dichotomies light/matter and continuous/discrete. Relativity and quantum mechanics, separately, brought us close to success, and the Dirac equation, with its implication of spin,

[d]In fact these particles obey wave equations that do have negative-energy solutions.

brought us closer still. But so far we haven't quite got there. Photons are evanescent, electrons ... well, they're evanescent too, as a matter of experimental fact, as I just mentioned, but we haven't yet adequately fit that feature into our theoretical discussion. In hole theory electrons can come and go, but only as positrons go and come.

These are not so much contradictions as indications of missed opportunity. They indicate that there ought to be some alternative to hole theory that covers all forms of matter, and that treats the creation and destruction of particles as a primary phenomenon.

Ironically, Dirac himself had earlier constructed the prototype of such a theory. In 1927, he applied the principles of the new quantum mechanics to Maxwell's equations of classical electrodynamics. He showed that Einstein's revolutionary postulate that light comes in particles – photons – was a consequence of the logical application of these principles, and that the properties of photons were correctly accounted for. Few observations are so common as that light can be created from non-light, say by a flashlight, or aborbed and annihilated, say by a black cat. But translated into the language of photons, this means that the quantum theory of Maxwell's equations is a theory of the creation and destruction of particles (photons). Indeed, the electromagnetic field appears, in Dirac's quantum theory of electromagnetism, primarily as an agent of creation and destruction. Photons arise as excitations of this field, which is the primary object. Photons come and go, but the field abides. The full significance of this development seems to have escaped Dirac and all of his contemporaries for some time, perhaps precisely because of the apparent specialness of light (dichotomy!). But it is a general construction, which can be applied to the object that appears in Dirac's equation – the electron field – as well.

The result of a logical application of the principles of quantum mechanics to Dirac's equation is an object similar to what he found for Maxwell's equations. It is an object that destroys electrons, and creates positrons[e]. Both are examples of *quantum fields.* When the object that appears in Dirac's equation is interpreted as a quantum field, the negative-energy solutions take on a completely different meaning, with no problematic aspects. The positive-energy solutions multiply electron destruction operators, while the negative-energy solutions multiply positron creation operators. In this framework, the difference between the two kinds of solution is that negative

[e]There is also a closely related object, the Hermitean conjugate, that creates electrons and destroys positrons.

energy represents the energy you need to borrow to make a positron, while positive energy is what you gain by destroying an electron. The possibility of negative numbers is no more paradoxical here than in your bank balance.

With the development of quantum field theory, the opportunities that Dirac's equation and hole theory made evident, but did not quite fulfill, were finally met. The description of light and matter was put, at last, on a common footing. Dirac said, with understandable satisfaction, that with the emergence of quantum electrodynamics physicists had attained foundational equations adequate to describe "all of chemistry, and most of physics".

In 1932 Enrico Fermi constructed a successful theory of radioactive decays (beta decays), including the neutron decay I mentioned before, by exporting the concepts of quantum field theory far from their origin. Since these processes involve the creation and destruction of protons – the epitome of 'stable' matter – the old dichotomies had finally been transcended. Both particles and light are epiphenomena, surface manifestations of the deeper and abiding realities, quantum fields. These fields fill all of space, and in this sense they are continuous. But the excitations they create, whether we recognize them as particles of matter or as particles of light, are discrete.

In hole theory we had a picture of the vacuum as filled with a sea of negative-energy electrons. In quantum field theory, the picture is quite different from this. But there is no returning to innocence. The new picture of the vacuum differs even more radically from naive "empty space". Quantum uncertainty, combined with the possibility of processes of creation and destruction, implies a vacuum teeming with activity. Pairs of particles and antiparticles fleetingly come to be and pass away. I once wrote a sonnet about virtual particles, and here it comes:

Beware of thinking nothing's there –
Remove what you can; despite your care
Behind remains a restless seething
Of mindless clones beyond conceiving.

They come in a wink, and dance about;
Whatever they touch is seized by doubt:
What am I doing here? What should I weigh?
Such thoughts often lead to rapid decay.

Fear not! The terminology's misleading;
Decay is virtual particle breeding

And seething, though mindless, can serve noble ends,
The clone-stuff, exchanged, makes a bond between friends.

To be or not? The choice seems clear enough,
But Hamlet oscillated. So does this stuff.

11. Aftermaths

With the genesis of quantum field theory, we reach a natural intellectual boundary for our discussion of the Dirac equation. By the mid-1930s the immediate paradoxes this equation raised had been resolved, and its initial promise had been amply fulfilled. Dirac received the Nobel Prize in 1933, Anderson in 1935.

In later years the understanding of quantum field theory deepened, and its applications broadened. Using it, physicists have constructed (and established with an astonishing degree of rigor and beyond all reasonable doubt) what will stand for the foreseeable future – perhaps for all time – as the working Theory of Matter. How this happened, and the nature of the theory, is an epic story involving many other ideas, in which the Dirac equation as such plays a distinguished but not a dominant role. But some later developments are so closely linked to our main themes, and so pretty in themselves, that they deserve mention here.

There is another sense in which the genesis of quantum field theory marks a natural boundary. It is the limit beyond which Dirac himself did not progress. Like Einstein, in his later years Dirac took a separate path. He paid no attention to most of the work of other physicists, and dissented from the rest. In the marvelous developments that his work commenced, Dirac's own participation was peripheral.

11.1. *QED and Magnetic Moments*

Interaction with the ever-present dynamical vacuum of quantum field theory modifies the observed properties of particles. We do not see the hypothetical properties of the "bare" particles, but rather the physical particles, "dressed" by their interaction with the quantum fluctuations in the dynamical vacuum.

In particular, the physical electron is not the bare electron, and it does not quite satisfy Dirac's $g = 2$. When Polykarp Kusch made very accurate measurements, in 1947, he found that g is larger than 2 by a factor 1.00119. Now this is not a very large correction, quantitatively, but it was a great stimulus to theoretical physics, because it provided a very concrete challenge. At that time there were so many loose ends in fundamental physics – a plethora of unexpected, newly discovered particles including muons, π mesons, and others, no satisfactory theory explaining what force holds atomic nuclei together, fragmentary and undigested results about radioactive decays, anomalies in high-energy cosmic rays – that it was hard to know where to focus. In fact, there was a basic philosophical conflict about strategy.

Most of the older generation, the founders of quantum theory, including Einstein, Schrödinger, Bohr, Heisenberg, and Pauli, were prepared for another revolution. They thought it was fruitless to spend time trying to carry out more accurate calculations in quantum electrodynamics, since this theory was surely incomplete and probably just wrong. It did not help that the calculations required to get more accurate results are very difficult, and that they seemed to give senseless (infinite) answers. So the old masters were searching for a different kind of theory, unfortunately with no clear direction.

Ironically, it was a younger generation of theorists – Schwinger, Feynman, Dyson, and Tomonaga in Japan – who played a conservative role[f]. They found a way to perform the more accurate calculations, and get meaningful finite results, without changing the underlying theory. The theory they used, in fact, was just the one Dirac had constructed in the 20s and 30s. The result of an epochal calculation by Schwinger, including the effects of the dynamic vacuum, was a small correction to Dirac's $g = 2$. It too was reported in 1947, and it agreed spectacularly well with Kusch's contemporary measurements. Many other triumphs followed. Kusch received the Nobel Prize in 1955; Schwinger, Feynman, and Tomonaga jointly in 1965 (the delay is hard to understand!).

Strangely enough, Dirac did not accept the new procedures. Caution was perhaps justified in the early days, when the mathematical methods being used were unfamiliar and not entirely well defined and involved a certain amount of inspired guesswork. But the technical difficulties were cleaned up in due course. [g]

[f]Seminal contributions were also made by the slightly older theorists Kramers and Bethe, and by the theorist-turned-experimentalist Lamb.

[g]Although QED does have problems of principle, if it is regarded (unrealistically!) as a completely closed theory, they are problems at a different level than what troubled Dirac, and they are very plausibly solved by embedding QED into a larger, asymptotically free theory - see below. This has very little practical effect on most of its predictions.

Feynman called QED "the jewel of physics – our proudest possession." But in 1951 Dirac wrote

> Recent work by Lamb, Schwinger and Feynman and others has been very successful... but the resulting theory is an ugly and incomplete one.

And in his last paper, in 1984,

> These rules of renormalization give surprisingly, excessively good agreement with experiments. Most physicists say that these working rules are, therefore, correct. I feel that this is not an adequate reason. Just because the results happen to be in agreement with experiment does not prove that one's theory is correct.

You might notice a certain contrast in tone between the young Dirac, who clung to his equation like a barnacle because it explained experimental results, and the older inhabitant of the same body.

Today the experimental determination of the magnetic moment of the electron is

$$(g/2)_{\text{experiment}} = 1.001\,159\,652\,188\,4\,(43)$$

while the theoretical prediction, firmly based on QED, calculated to high accuracy, is

$$(g/2)_{\text{theory}} = 1.001\,159\,652\,187\,9\,(43)$$

where the uncertainty in the last two digits is indicated. It is the toughest, most accurate confrontation between intricate – but precisely defined! – theoretical calculations and delicate – but precisely controlled! – experiments in all of science. That's what Feynman meant by "our proudest possession".

Ever more accurate determination of the magnetic moment of the electron, and of its kindred particle the muon, remains an important frontier of experimental physics. With the accuracies now achievable, the results will be sensitive to effects of quantum fluctuations due to hypothetical new heavy particles – in particular, those expected to be associated with supersymmetry.

11.2. *QCD and the Theory of Matter*

The magnetic moment of the proton does not satisfy Dirac's $g = 2$, but instead has $g \approx 5.6$. For neutrons it is worse. Neutrons are electrically neutral, so the simple Dirac equation for neutrons predicts no magnetic moment at all. In fact the neutron has a magnetic moment about 2/3 as large as that of a proton, and with the opposite orientation relative to spin. That corresponds to an infinite value of g, since the neutron is electrically neutral. The discrepant values of these magnetic moments were the earliest definite indication that protons and neutrons are more complicated objects than electrons.

With further study, many more complications appeared. The forces among protons and neutrons were found to be very complicated. They depend not only on the distance between them, but also on their velocities, and spin orientations, and all combinations of these together, in a bewildering way. In fact, it soon appeared that they are not "forces" in the traditional sense at all. To have a force between protons, in the traditional sense, would mean that the motion of one proton can be affected by the presence of another, so that when you shoot one proton by another, it swerves. What you actually observe is that when one proton collides with another, typically many particles emerge, most of which are highly unstable. There are π mesons, K mesons, ρ mesons, Λ and Σ baryons, their antiparticles, and many more. All these particles interact very powerfully with each other. And so the problem of nuclear forces, a frontier of physics starting in the 1930s, became the problem of understanding a vast new world of particles and reactions, the most powerful in Nature. Even the terminology changed. Physicists no longer refer to nuclear forces, but to the strong interaction.

Now we know that all the complexities of the strong interaction can be described, at a fundamental level, by a theory called quantum chromodynamics, or QCD, a vast generalization of QED. The elementary building blocks of QCD are quarks and gluons. There are six different kinds, or 'flavors', of quarks: u, d, s, c, b, t (up, down, strange, charm, bottom, top). The quarks are very similar to one another, differing mainly in their mass. Only the lightest ones, u and d, are found in ordinary matter. Making an analogy to the building blocks of QED, quarks play roughly the role of electrons, and gluons play roughly the role of photons. The big difference is that whereas in QED there is just one type of charge, and one photon, in QCD there are three types of charge, called colors, and eight gluons. Some gluons respond to color charges, similarly to the way photons respond to electric charge. Others mediate transitions between one color and another. Thus (say) a u quark with blue charge can radiate a gluon and turn into a u quark with green charge. Since all the charges overall must be conserved, this particular gluon must have blue charge +1, green charge −1. Since gluons themselves carry unbalanced color charge, in QCD there are elementary processes where gluons radiate other gluons. There is nothing like this in QED. Photons are electrically neutral, and to a very good approximation they do not interact with other photons. Much of the richness and complexity of QCD arises because of this new feature.

Described thus baldly and verbally, without grounding in concepts or phenomena, QCD might seem both arbitrary and fantastic. In fact QCD is a theory of compelling symmetry and mathematical beauty. Unfortunately, I won't be able to do justice to those aspects here. But some brief explications are in order .···

How did we arrive at such a theory? And how do we know it's right? In the case of QCD, these are two very different questions. The historical path to its discovery was tortuous, with many false trails and blind alleys. But in retrospect, it didn't have to be that way. If the right kind of ultra-high-energy accelerators had come on line earlier, QCD would have stared us in the face[h]. This *gedanken*-history brings together most of the ideas I've discussed in this article, and forms a fitting conclusion to its physical part.

When electrons and positrons are accelerated to ultrahigh energy and then made to collide, two kinds of events are observed. In one kind of event the particles in the final states are leptons and photons. For this class of events, usually the final state is just a lepton and its anti-lepton; but in about 1% of the events there is also a photon, and in about 0.01% of the events there are also two photons. The probability for these sorts of events, and for the various particles to come out at various angles with different energies, can all be computed using QED, and it all works out very nicely. Conversely, if you hadn't known about QED, you could have figured out the basic rules for the fundamental interaction of QED – that is, the emission of a photon by an electron – just by studying these events. The fundamental interaction of light with matter is laid out right before your eyes.

In the other kind of event, you see something rather different. Instead of just two or at most a handful of particles coming out, there are many. And they are different kinds of particles. The particles you see in this second class of events are things like π mesons, K mesons, protons, neutrons, and their antiparticles – all particles that, unlike photons and leptons, have strong interactions. The angular distribution of these particles is very structured. They do not come out independently, every which way. Rather, they emerge

---

[h]Up to a couple of profound but well-posed and solvable problems, as I'll shortly discuss.

in just a few directions, making narrow sprays or (as they're usually called) "jets". About 90% of the time there are just two jets, in opposite directions; roughly 10% of the time there are three jets, 1% four jets – you can guess the pattern.

Now if you squint a little, and don't resolve the individual particles, but just follow the flow of energy and momentum, then the two kinds of events – the QED 'particle' events, and the 'jetty' events with strongly interacting particles – look just the same!

So (in this imaginary history) it would have been hard to resist the temptation to treat the jets as if they are particles, and propose rules for the likelihood of different radiation patterns, with different numbers, angles, and energies of the jet-particles, in direct analogy to the procedures that work for QED. And this would work out very nicely, because rules quite similar to those for QED actually do describe the observations. Of course, the rules that work are precisely those of QCD, including the new processes where glue radiates glue. All these rules – the foundational elements of the entire theory – could have been derived directly from the data. "Quarks" and "gluons" would be words with direct and precise operational definitions, in terms of jets.

Still, there would have been two big conceptual puzzles. Why do the experiments show 'quarks' and 'gluons' instead of just quarks and gluons – that is, jets, instead of just particles? And how do you connect the theoretical concepts that directly and successfully describe the high-energy events to all the other phenomena of the strong interaction? The connection between the supposedly foundational theory and the mundane observations is, to say the least, not obvious. For example, you would like to construct protons out of the 'quarks' and 'gluons' that appear in the fundamental theory. But this looks hopeless, since the jets in terms of which 'quarks' and 'gluons' are operationally defined often contain, among other things, protons.

There is an elegant solution to these problems. It is the phenomenon of *asymptotic freedom* in QCD. According to asymptotic freedom, radiation events that involve large changes in the flow of energy and momentum are rare, while radiation events that involve only small changes in energy and momentum are very common. Asymptotic freedom is not a separate assumption, but a deep mathematical consequence of the structure of QCD.

Asymptotic freedom neatly explains why there are jets in electron-positron annihilations at high energies, in the class of events containing strongly interacting particles. Immediately after the electron and positron annihilate, you have a quark and an antiquark emerging. They are moving rapidly, in opposite directions. They quickly radiate gluons, and the gluons themselves radiate, and a complicated cascade develops, with many particles. But despite all this commotion the overall flow of energy and momentum is not significantly disturbed. Radiations that disturb the flow of energy and momentum are rare, according to asymptotic freedom. So there is a large multiplicity of particles all moving in the same direction, the direction originally staked out by the quark or antiquark. In a word, we've produced a jet. When one of those rare radiations that disturbs the flow of energy and momentum takes place, the radiated gluon starts a jet of its own. Then we have a three-jet event. And so forth.

Asymptotic freedom also indicates why the description of protons (and the other strongly interacting particles) that we actually observe as individual stable, or quasi-stable, entities are complicated objects. For such particles are, more or less by definition, configurations of quarks, antiquarks, and gluons that have a reasonable degree of stability. But since the quarks, antiquarks, and gluons all have a very high probability for radiating, no simple configuration will have this property. The only possibility for stability involves dynamic equilibrium, in which the emission of radiation in one part of the system is balanced by its absorption somewhere else.

As things actually happened, asymptotic freedom was discovered theoretically (by David Gross and me, and independently by David Politzer) and QCD was proposed as the theory of the strong interaction (by Gross and me) in 1973, based on much less direct evidence. The existence of jets was anticipated, and their properties were *predicted* theoretically, in considerable detail, before their experimental observation. Based on these experiments, and many others, today QCD is accepted as the fundamental theory of the strong interaction, on a par with QED as the description of the electromagnetic interaction.

There has also been enormous progress in using QCD to describe the properties of protons, neutrons, and the other strongly interacting particles. This involves very demanding numerical work, using the most powerful computers, but the results are worth it. One highlight is that we can calculate from first principles, with no important free parameters, the masses of protons and neutrons. As I explained, from a fundamental point of view these particles are quite complicated dynamical equilibria of quarks, antiquarks, and gluons. Most of their mass – and therefore most of the mass of matter, including human brains and bodies – arises from the pure energy of these objects, themselves essentially massless, in motion, according to $m = E/c^2$. At this level, at least, we are ethereal creatures.

Dirac said that QED described "most of physics, and all of chemistry". Indeed, it is the fundamental theory of the outer structure of atoms (and much more). In the same sense, QCD is the fundamental theory of atomic nuclei (and much more). Together, they constitute a remarkably complete, well tested, fruitful and economical Theory of Matter.

12. The Fertility of Reason

I've now discussed in some detail how "playing with equations" led Dirac to an equation laden with consequences that he did not anticipate, and that in many ways he resisted, but that proved to be true and enormously fruitful. How could such a thing happen? Can mathematics be truly creative? Is it really possible, by logical processing or calculation, to arrive at essentially new insights – to get out more than you put in?

This question is especially timely today, since it lies at the heart of debates regarding the nature of machine intelligence – whether it may develop into a species of mind on a par with human intelligence, or even its eventual superior.

At first sight, the arguments against appear compelling.

Most powerful, at least psychologically, is the argument from introspection. Reflecting on our own thought processes, we can hardly avoid an unshakeable intuition that they do not consist exclusively, or even primarily, of rule-based symbol manipulation. It just doesn't feel that way. We normally think in images and emotions, not just symbols. And our streams of thought are constantly stimulated and redirected by interactions with the external world, and by internal drives, in ways that don't seem to resemble at all the unfolding of mathematical algorithms.

Another argument derives from our experience with modern digital computers. For these are, in a sense, ideal mathematicians. They follow precise rules (axioms) with a relentlessness, speed, and freedom from error that far surpasses what is possible for humans. And in many specialized, essentially mathematical tasks, such as arranging airline flight or oil delivery schedules to maximize profits, they far surpass human performance. Yet by common, reasonable standards even the most powerful modern computers remain fragile, limited, and just plain dopey. A trivial programming mistake, a few lines of virus code, or a memory flaw can bring a powerful machine to a halt, or send it into an orgy of self-destruction. Communication can take place only in a rigidly controlled format, supporting none of the richness of natural language. Absurd output can, and often does, emerge uncensored and unremarked.

Upon closer scrutiny, however, these arguments raise questions and doubts. Although the nature of the map from patterns of electrical signals in nerve cells to processes of human thought remains deeply mysterious in many respects, quite a bit is known, especially about the early stages of sensory processing. Nothing that has been discovered so far suggests that anything more exotic than electric and chemical signalling, following well-established physical laws, is involved. The vast majority of scientists accept as a working hypothesis that a map from patterns of electric signals to thought must and does exist. The pattern of photons impinging on our retina is broken up and parsed out into elementary units, fed into a bewildering series of different channels, processed, and (somehow) reassembled to give us the deceptively simple "picture of the world", organized into objects in space, that we easily take for granted. The fact is we do not have the slightest idea how we accomplish most of what we do, even – perhaps especially – our most basic mental feats. People who've attempted to construct machines that can recognize objects appearing in pictures, or that can walk around and explore the world like a toddler, have had a very frustrating time, even though they can do these things very easily themselves. They can't teach others how they do these things because they don't know themselves. Thus it seems clear that introspection is an unreliable guide to the deep structure of thought, both as regards what is known and what is unknown.

Turning to experience with computers, any negative verdict is surely premature, since they are evolving rapidly. One recent benchmark is the victory of Deep Blue over the great world chess champion Garry Kasparov in a brief match. No one competent to judge would deny that play at this level would be judged a profoundly creative accomplishment, if it were performed by a human. Yet such success in a limited domain only sharpens the question: What is missing, that prevents the emergence of creativity from pure calculation over a broad front? In thinking about this tremendous question, I believe case studies can be of considerable value.

In modern physics, and perhaps in the whole of intellectual history, no episode better illustrates the profoundly creative nature of mathematical reasoning than the history of the Dirac equation. In hindsight, we know that what Dirac was trying to do is strictly impossible. The rules of quantum mechanics, as they were understood in 1928, cannot be made consistent with special relativity. Yet from inconsistent assumptions Dirac was led to an equation that remains a cornerstone of physics to this day.

So here we are presented with a specific, significant, well-documented example of how mathematical reasoning about the physical world, culminating in a specific equation, led to results that came as a complete surprise to the thinker himself. Seemingly in defiance of some law of conservation, he got out much more than he put in. How was such a leap possible? Why did Dirac, in particular, achieve it? What drove Dirac and his contemporaries to persist in clinging to his equation, when it led them out to sea?[i]

Insights emerge from two of Dirac's own remarks. In his characteristically terse essay "My Life as a Physicist" he pays extended tribute to the value of his training as an engineer, including:

> The engineering course influenced me very strongly.··· I've learned that, in the description of nature, one has to tolerate approximations, and that even work with approximations can be interesting and can sometimes be beautiful.

Along this line, one source of Dirac's (and others') early faith in his equation, which allowed him to overlook its apparent flaws, was simply that he could find approximate solutions of it that agreed brilliantly with experimental data on the spectrum of hydrogen. In his earliest papers he was content to mention, without claiming to solve, the difficulty that there were other solutions, apparently equally valid mathematically, that had no reasonable physical interpretation.

Along what might superficially seem to be a very different line, Dirac often paid tribute to the heuristic power of mathematical beauty:

> The research worker, in his efforts to express the fundamental laws of Nature in mathematical form, should strive mainly for mathematical beauty.

This was another source of early faith in Dirac's equation. It was (and is) extraordinarily beautiful.

Unfortunately, it is difficult to make precise, and all but impossible to convey to a lay reader, the nature of mathematical beauty. But we can draw some analogies with other sorts of beauty. One feature that can make

[i]Much later, in the 1960s, Heisenberg recalled "Up till that time [1928] I had the impression that, in quantum theory, we had come into the harbor, into the port. Dirac's paper threw us out into the sea again."

a piece of music, a novel, or a play beautiful is the accumulation of tension between important, well-developed themes, which is then resolved in a surprising and convincing way. One feature that can make a work of architecture or sculpture beautiful is symmetry – balance of proportions, intricacy toward a purpose. The Dirac equation possesses both these features to the highest degree.

Recall that Dirac was working to reconcile the quantum mechanics of electrons with special relativity. It is quite beautiful to see how the tension between conflicting demands of simplicity and relativity can be harmonized, and to find that there is essentially only one way to do it. That is one aspect of the mathematical beauty of the Dirac equation. Another aspect, its symmetry and balance, is almost sensual. Space and time, energy and momentum, appear on an equal footing. The different terms in the system of equations must be choreographed to the music of relativity, and the pattern of 0s and 1s (and i s) dances before your eyes.

The lines converge when the needs of physics lead to mathematical beauty, or – in rare and magical moments – when the requirements of mathematics lead to physical truth. Dirac searched for a mathematical equation satisfying physically motivated hypotheses. He found that to do so he actually needed a system of equations, with four components. This was a surprise. Two components were most welcome, as they clearly represented the two possible directions of an electron's spin. But the extra doubling at first had no convincing physical interpretation. Indeed, it undermined the assumed meaning of the equation. Yet the equation had taken on a life of its own, transcending the ideas that gave birth to it, and before very long the two extra components were recognized to portend the spinning positron, as we saw.

With this convergence, I think, we reach the heart of Dirac's method in reaching the Dirac equation, which was likewise Maxwell's in reaching the Maxwell equations, and Einstein's in reaching both the special and the general theories of relativity. They proceed by *experimental logic.* That concept is an oxymoron only on the surface. In experimental logic, one formulates hypotheses in equations, and experiments with those equations. That is, one tries to improve the equations from the point of view of beauty and consistency, and then checks whether the "improved" equations elucidate some feature of Nature. Mathematicians recognize the technique of "proof by contradiction": To prove A, you assume the opposite of A, and reach a contradiction. Experimental logic is "validation by fruitfulness": To validate A, assume it, and show that it leads to fruitful consequences. Relative to routine deductive logic, experimental logic abides by the Jesuit credo "It is more blessed to ask forgiveness than permission." Indeed, as we have seen, experimental logic does not regard inconsistency as an irremediable catastrophe. If a line of investigation has some success, and is fruitful, it should not be abandoned on account of its inconsistency, or its approximate nature. Rather, we should look for a way to make it true.

With all this in mind, let us return to the question of the creativity of mathematical reasoning. I said before that modern digital computers are, in a sense, ideal mathematicians. Within any reasonable, precisely axiomatized domain of mathematics, we know how to program a computer so it will systematically prove all the valid theorems[j]. A modern machine of this sort could churn through its program, and output valid theorems, much faster and more reliably than any human mathematician could. But running such a program to do advanced mathematics would be no better than setting the proverbial horde of monkeys to typing, hoping to reproduce Shakespeare. You'd get a lot of true theorems, but essentially all of them would be trivial, with the gems hopelessly buried amidst the rubbish. In practice, if you peruse journals of mathematics or mathematical physics, not to speak of literary magazines, you won't find much work submitted by computers. Attempts to teach computers to do "real" creative mathematics, like the attempts to teach them to recognize real objects or navigate the real world, have had very limited success. Now we begin to see that these are closely related problems. Creative mathematics and physics rely not on perfect logic, but rather on an experimental logic. Experimental logic involves noticing patterns, playing with them, making assumptions to explain them, and – especially – recognizing beauty. And creative physics requires more: abilities to sense and cherish patterns in the world, and to value not only logical consistency, but also (approximate!) fidelity to the world as observed.

So, returning to the central question: Can purely mathematical reasoning be creative? Undoubtedly, if it is used a la Dirac, in concert with the

[j]This is a consequence of Godel's completeness theorem for first-order predicate logic. Sophisticated readers may wonder how this result, that all valid theorems can be proved in mechanical fashion, can be consistent with Godel's famous incompleteness theorem. (It's not a misprint: Godel proved both completeness and incompleteness theorems.) To make a long story short, Godel's incompleteness theorem shows that in any rich mathematical system you will be able to formulate meaningful statements such that neither the statement nor its denial is a theorem. Such "incompleteness" does not contradict the possibility of systematically enumerating all the theorems.

abilities to tolerate approximations, to recognize beauty, and to learn by interacting with the real world. Each of these factors has played a role in all the great episodes of progress in physics. The question returns, as a challenge to ground those abilities in specific mechanisms.

Acknowledgments

My work is supported in part by funds provided by the U.S. Department of Energy (D.O.E.) under cooperative research agreement #DF-FC02-94ER40818. This presentation is adapted from my chapter "A Piece of Magic: The Dirac Equation" in the book *It Must Be Beautiful, The Great Equations of Modern Science*, ed. G. Farmelo (Granta Books, 2002).

References

1. For background material on atomic physics and quantum theory, including excerpts from important original sources, I highly recommend H. Boorse and L. Motz, *The World of the Atom* (Basic Books, 1966). Of course, some of its more "timely" parts appear somewhat dated today.
2. Dirac's classic is *The Principles of Quantum Mechanics* (Fourth Edition, Cambridge 1958).
3. A demanding but honest and beautiful treatment of the principles of quantum electrodynamics, with no mathematical prerequisites, is R.P. Feynman, *QED: The Strange Theory of Light and Matter* (Princeton. 1985).
4. For a brief account of QCD, easily accessible after Feynman's book, with no mathematical prerequisites, see F. Wilczek, "QCD Made Simple", *Physics Today*, **53N8** 22–28, (2000) . I'm at work on a full account, to be called simply *QCD* (Princeton).
5. For a conceptual review of quantum field theory, see my article "Quantum Field Theory" in the American Physical Society Centenary issue of *Rev. Mod. Phys.* **71**, S85–S95, (1999); this issue is also published as *More Things in Heaven and Earth – A Celebration of Physics at the Millennium*, B. Bederson, ed. (Springer-Verlag, New York), (1999) It contains several other reflective articles that touch on many of our themes.

William J. Marciano

Proceedings of the Dirac Centennial Symposium
Howard Baer and Alexander Belyaev

Anomalous Magnetic Moments

William J. Marciano
Physics Department
Brookhaven National Laboratory
Upton, NY 11973, USA

The Dirac equation explained why the gyromagnetic ratio, g factor, is equal to 2 for fundamental spin $\frac{1}{2}$ particles. Quantum loop effects were subsequently shown to induce a small shift or anomaly, $a \equiv (g-2)/2$. Anomalous magnetic moment effects have been calculated and measured with extraordinary precision for the electron and muon. Here, the Standard Model's predictions for $a_l = (g_l - 2)/2$, $l = e, \mu$ are described and compared with experimental values. Implications for probing "New Physics" effects are also discussed.

13. The Dirac Equation and $g = 2$

In 1928, at the age of 25, Dirac introduced[1] his now famous equation that described a 4 component (spinor) electron wavefunction, $\psi(x)$, in an electromagnetic potential, $A_\mu(x)$:

$$i(\partial_\mu - ie\, A_\mu(x))\gamma^\mu \psi(x) = m_e \psi(x) \tag{1}$$

where the γ^μ are 4x4 Dirac matrices. That simple equation elegantly combined quantum mechanics, special relativity and spin with the principle of electromagnetic gauge invariance. It has become one of the cornerstones of Modern Physics, providing a foundation for Quantum Electrodynamics (QED) and the entire Standard Model paradigm of elementary particle physics.

The Dirac equation is celebrated for its revolutionary prediction that antiparticles must exist as negative energy solutions; however, that revelation was secondary. Its immediate success was in providing an explanation for why the gyromagnetic ratio, g_e, that parametrizes the relationship between the electron's magnetic moment, $\boldsymbol{\mu}_e$, and its intrinsic spin, $\boldsymbol{S}$,

$$\boldsymbol{\mu}_e = g_e \frac{e}{2m_e} \boldsymbol{S}\,, \tag{2}$$

is equal to 2. If spin and the intrinsic magnetic moment were related in the same way as orbital angular momentum, $\boldsymbol{L}$, and its associated magnetic moment, one would expect $g_e = 1$. However, the empirical requirement that $g_e = 2$ was already well established by atomic spectroscopy in 1928. The Dirac equation provided a natural theoretical underpinning for that value.[2]

One can discern $g_e = 2$ by applying the operator $(\partial_\nu - ie\, A_\nu(x))\gamma^\nu$ to eq. (1). After some manipulation one finds that each component of $\psi(x)$ obeys a scalar Klein-Gordon equation but there is an additional term describing a magnetic moment with $g_e = 2$ interacting with a magnetic field.[2]

It is interesting to note that fundamental spin one non-Abelian (Yang-Mills) gauge bosons also have $g = 2$ rather than 1. That special value is deeply connected with their renormalizability and other remarkable properties (e.g. asymptotic freedom in the case of Quantum Chromodynamics (QCD)).

Of course, Dirac could have accommodated large or small deviations from $g_e = 2$ by adding a so-called Pauli interaction term[3]

$$-\tfrac{e}{2m_e} a_e F_{\mu\nu}(x)\sigma^{\mu\nu}\psi(x) \tag{3}$$

$$F_{\mu\nu} = \partial_\mu A_\nu - \partial_\nu A_\mu \tag{4}$$

$$\sigma^{\mu\nu} = \tfrac{i}{2}[\gamma_\mu, \gamma_\nu] \tag{5}$$

to eq. (1). Then one would find

$$g_e = 2(1 + a_e) \tag{6}$$

where a_e would give rise to an arbitrary or anomalous deviation from the simple Dirac value. Such a term is very much phenomenologically required for the proton and neutron, where large deviations[4] from 2

$$g_p \simeq 5.59$$
$$g_n \simeq -3.83 \tag{7}$$

steming from their composite structure were later observed.

What forbids (at least in lowest order) the addition of a Pauli term in the case of elementary fermions such as the electron and muon? Such a term preserves Lorentz covariance and local gauge invariance. Dirac probably excluded it because of his guiding principles of simplicity and elegance as well as his use of minimal coupling, i.e. replacing ∂_μ by the covariant derivative $\partial_\mu - ie\,A_\mu$. Of course, given the experimental evidence for $g_e = 2$, there was no need for a Pauli term.

Today, we would also automatically exclude Pauli interactions in our fundamental Lagrangian because they correspond to dimension 5 operators which spoil renormalizability. However, given Dirac's disdain for (infinite) renormalization, he would probably not find that argument compelling. Another (more exotic) way to forbid such a term is to require supersymmetry at a fundamental Lagrangian level.[5] That enlargement of Poincare invariance to include extra spinorial generators (a very Dirac like idea) would link the electron with its scalar partners (selectrons) and forbid fundamental Pauli terms. I mention that connection because, as we shall see, the muon anomalous magnetic measurement with high precision may provide a window to (broken) supersymmetry effects at the quantum loop level. Also, some extended supersymmetric theories are finite, a feature that would have appealed to Dirac.

Whatever Dirac's reason for excluding Pauli terms (most likely their lack of observation), it is clear that his explanation for why $g_e = 2$ clarified an outstanding problem in atomic spectroscopy. It represented a great triumph for the Dirac equation, but was not the end of the magnetic moment story.

14. The Electron Anomalous Magnetic Moment

In 1947, (nearly 20 years after Dirac's equation was introduced), small anomalous effects began to be observed[6] in precision studies of hyperfine spectroscopy in hydrogen and deuterium (at the 0.1% level). Gregory Breit suggested [7] on empirical grounds that the small shifts could be explained if g_e deviated slightly from 2. Schwinger then demonstrated[8] the power of QED by computing the predicted quantum loop contribution to a_e

$$a_e \equiv \frac{g_e - 2}{2} = \frac{\alpha}{2\pi} \simeq 0.00112. \tag{8}$$

That finite, unambiguous result was in good agreement with experiment. It represented a tremendous quantitative triumph for QED and quantum field theory. It also ushered in an era of very precise experimental measurements which were used to test the validity of QED to many significant figures and search for deviations stemming from "New Physics" effects. It is interesting to note that a rather novel derivation (using a Hamiltonian formalism) of Schwinger's result in eq. (8) was given by Dirac in his 1963-64 lectures on quantum field theory.[9] Today, computation of $a_e = \alpha/2\pi$ is a basic exercise in essentially all quantum field theory courses.[10] It is made relatively simple by Feynman diagram techniques.

Currently, the most precise measurements of a_e for the electron and positron (using Penning traps) by H. Dehmeldt and collaborators give[4]

$$\begin{aligned} a_{e-}^{exp} &= 11596521884(43) \times 10^{-13} \\ a_{e+}^{exp} &= 11596521879(43) \times 10^{-13}, \end{aligned} \tag{9}$$

where the numbers in parenthesis represent one sigma uncertainties in the last two decimal places. The number of significant digits in those numbers is impressive, a truly spectacular achievement by any standard. A new experiment[11] now in progress at Harvard aims to further reduce the errors in eq. (9) by a factor of 15, significant improvement.

The agreement between a_{e-}^{exp} and a_{e+}^{exp} provides a sensitive test of CPT symmetry which requires they be identical. Those very precise values can also be compared with the theoretical prediction which has been computed in QED through 4 loops[12]

$$a_e = \frac{\alpha}{2\pi} - 0.328478444(\frac{\alpha}{\pi})^2 + 1.181234(\frac{\alpha}{\pi})^3 - 1.7502(\frac{\alpha}{\pi})^4 + 1.70 \times 10^{-12}. \tag{10}$$

Some comments about that result are in order:

- Muon and tau vacuum polarization loop effects[13] are included in the $O(\frac{\alpha}{\pi})^2$ and $O(\frac{\alpha}{\pi})^3$ coefficients (sometimes such effects are treated separately).
- The $O(\frac{\alpha}{\pi})^4$ coefficient has recently been revised.[12]
- The 1.70×10^{-12} contribution stems from 2 and 3 loop hadronic effects (1.67×10^{-12}) as well as very small 1 and 2 loop electroweak effects (0.03×10^{-12}). Such contributions will only start to be probed by the next generation (Harvard)measurement of a_e.

The perturbative expansion in $\frac{\alpha}{\pi}$ for a_e is very well behaved, with alternating coefficients of order 1. To compare experiment in eq. (9) with theory in eq. (10) requires a separate very precise direct determination of the fine structure constant, α. Currently, the quantum Hall effect gives[14]

$$\alpha^{-1} = 137.03600300(270), \tag{11}$$

which leads to the Standard Model (QED dominated) prediction

$$a_e^{SM} = 11596521524(230) \times 10^{-13}. \tag{12}$$

Other determinations[14] of α are less precise than eq. (11).

The prediction in eq. (12) is in relatively good agreement with the experimental values in eq. (9). That agreement represents one of the best tests of QED and perturbation theory. It is not a generally good probe of "New Physics" effects which are expected to be of the form[15]

$$\Delta a_e(NewPhysics) \simeq C(\frac{m_e}{\lambda})^2, \tag{13}$$

where $\wedge$ is the scale of New Physics and C is at most $O(1)$, but could be much smaller. The good agreement between theory and experiment gives for $C \simeq 1$ the constraint $\wedge > 80\ GeV$ which is not very prohibitive. Note, if Δa_e(New Physics) were linear in $m_e/\wedge$ rather than quadratic, one would get the very constraining bound $\wedge > 10^7 GeV$. However, that scenario is unrealistic. Because anomalous magnetic moments change chirality, like mass terms, one expects $\frac{1}{m_e}\Delta a_e$ to vanish as $m_e \to 0$.

Since "New Physics" is not yet likely to be affecting a_e^{exp}, one can use the comparison of eqs. (9) and (10) to determine α

$$\alpha^{-1}(a_e) = 137.03599877(40), \tag{14}$$

which is considerably better than any current direct determination ($Cf\ eq.$(11)).

The ongoing new measurement of a_e will reduce the error in eq. (9) and (14) by a factor of 15, a significant improvement. It will provide a sensitive test of CPT; however, to fully utilize it as a probe of "New Physics" will require a separate determination of α with comparable precision. Such an advance will be difficult; but might be best accomplished using the Rydberg constant in conjunction with a very precise m_e determination.[16]

15. The Muon Anomalous Magnetic Moment

Since "New Physics" contributions to a_l are expected to scale as $m_l^2/\wedge^2$, the muon anomalous magnetic moment, $a_\mu \equiv (g_\mu - 2)/2$, should be approximately $(m_\mu/m_e)^2 \simeq 40{,}000$ times more sensitive to "New Physics" than a_e. However the experimental a_μ is only about 200 times less precise than a_e, making a_μ overall about 200 times better for probing "New Physics". Of course, hadronic and electroweak loop contributions to a_μ^{SM} are also about 40,000 times larger than their currently negligible effect on a_e. So, a comparison of a_μ^{exp} and a_μ^{SM} must confront hadronic and electroweak loops with high precision.

On the experimental front, a series of experiments at CERN in the 1970s pushed a_μ^{exp} to about the $\pm 840 \times 10^{-11}$ level. That effort was more recently taken up again by experiment E821 at Brookhaven National Lab which has reported[17]

$$a_{\mu+}^{exp} = 116592030(80) \times 10^{-11}. \tag{15}$$

That value is already about a factor of 10 better than the classic CERN results. Data currently under analysis for $a_{\mu-}^{exp}$ will provide nearly as good precision, such that averaging $a_{\mu+}^{exp}$ and $a_{\mu-}^{exp}$ (assuming CPT) should soon lead to an overall uncertainty in a_μ^{exp} of about $\pm 60-65 \times 10^{-11}$. If funded for additional running, that experiment could probably be pushed to $\pm 20-30 \times 10^{-11}$ before systematic errors would start to dominate. In the much longer term, one could imagine a completely new experiment reaching $\pm 5 \times 10^{-11}$ with improved statistics and systematics.[18]

The ability to push a_μ^{exp} so far stems from some very special (unique) properties of the muon:

- Muons are copiously produced via $p + target \to \pi \to \mu\nu$. They are 100% polarized, but some polarization is lost during collection.
- The relatively long muon lifetime, $\tau_\mu \simeq 2.2 \times 10^{-6}$ sec, allows them to be studied in storage rings.
- The decay $\mu \to e\nu\bar{\nu}$ maximally violates parity. That allows one to monitor the muon polarization via the angle of the outgoing electron relative to the muon momentum.
- In an external magnetic field, a relativistic muon will precess as $sin(\frac{e}{2m_\mu} a_\mu Bt)$; so, one measures a_μ directly rather than g_μ.

At the magic $E_\mu \simeq 2.9 GeV$ value, lab electric fields do not give rise to additional background precession. Of course measuring a_μ^{exp} to better than 1 part per million requires similar or better precision on B and t. Fortunately, the technology for such precision exists.

The theoretical calculation of a_μ^{SM} represents a heroic tour de force which is generally divided into 3 parts

$$a_\mu^{sm} = a_\mu^{QED} + a_\mu^{Hadronic} + a_\mu^{EW}. \tag{16}$$

The QED contribution has been computed through 4 loops and roughly estimated at the 5 loop level (including lepton vacuum polarizations)[18]

$$\begin{aligned} a_\mu^{QED} = & \frac{\alpha}{2\pi} + 0.765857376(\frac{\alpha}{\pi})^2 + 24.05050898(\frac{\alpha}{\pi})^3 \\ & + 130(\frac{\alpha}{\pi})^4 + 930(\frac{\alpha}{\pi})^5. \end{aligned} \tag{17}$$

The coefficients are positive and growing. The coefficient of the $(\frac{\alpha}{\pi})^4$ term (and accordingly $(\frac{\alpha}{\pi})^5$) is currently being revised.[19] The value given, 130, should be taken as a temporary estimate that will be replaced by a final precise number in a short time. Using the value of α in eq. (14), one finds

$$a_\mu^{QED} = 116584728(10) \times 10^{-11}. \tag{18}$$

The rough uncertainty of $\pm 10 \times 10^{-11}$ given there is conservative. It should be reduced to about $\pm 2 \times 10^{-11}$ when the 4 loop revisions are fully completed. Even now, the uncertainty in eq. (18) is relatively small and does not affect the comparison of theory and experiment appreciably.

Hadronic loop corrections due to strongly interacting quarks and gluons start to contribute to a_μ^{SM} at the 2 loop level via hadronic vacuum polarization. A precise first principles QCD calculation of that effect is challenging, but may one day be possible using lattice gauge theory [20]; however, for now it is evaluated using data from $O(e^+e^- \to hadrons)$ and a dispersion relation. That approach currently gives[21]

$$a_\mu^{Hadronic}(vac.pol.) = 6847(70) \times 10^{-11}. \tag{19}$$

However, it is expected to shift upward due to anticipated normalization changes in some of the key $e^+e^- \to \pi^+\pi^-$ data coming from Novosibirsk. Indeed, an indication that a problem with eq. (19) existed was signaled by $\tau \to \nu_\tau +$ hadrons data which can be related by isospin to $e^+e^- \to$ hadrons in the dominant $I = 1$ channel. Using τ data, where available, in place of e^+e^- data and making isospin corrections gives[21]

$$a_\mu^{Hadronic}(vac.pol.) = 7090(50) \times 10^{-11} \quad (tau\ data). \tag{20}$$

Eqs. (19) and (20) are inconsistent, indicating a problem with the e^+e^- or τ data (perhaps both). Both methods are being reexamined. It is expected that they will move toward one another, the e^+e^- data due to luminosity changes and τ data due to refinements in the isospin corrections. For now, it seems prudent to average eqs. (19) and (20) while expanding the overall uncertainty (until consistency is clarified)

$$a_\mu^{Hadronic}(vac.pol.) = 6989(100) \times 10^{-11} \quad (Average). \tag{21}$$

Three loop hadronic effects have had a tumultuous history, having undergone significant changes over the years. Their estimated contribution now seems stable[22]

$$a_\mu^{Hadronic}(3loops) = -14(35) \times 10^{-11}, \tag{22}$$

with the uncertainty dominated by light by light hadronic diagrams. It will be a challenge to significantly reduce that error; but, such challenges are necessary to advance computational capabilities. In total, using eqs. (21) and (22)

$$a_\mu^{Hadronic} = 6975(106) \times 10^{-11}. \tag{23}$$

When the e^+e^- and τ data are clarified (and if consistent), the total hadronic error should be reduced to about $\pm 60 \times 10^{-11}$. Further reduction in the error should be possible with additional e^+e^- data using the radiative return process $e^+e^- \to \gamma$ + hadrons at high energy and luminosity colliders.[23]

Electroweak radiative corrections to a_μ^{SM} are on a very firm footing. They have been computed at the one[24] and two loop levels.[25] Indeed, the electroweak corrections to a_μ represent the first essentially complete 2 loop calculation in the Standard Model.[26] Various interesting features appear at the 2 loop level. For example, fermion triangle diagrams give divergent results due to the Adler- Bell- Jackiw anomaly. Divergences from leptons cancel with those from quarks and leave a finite contribution. Also, the Higgs scalar gives a much larger 2 loop contribution than its (completely negligible) one loop effect. Two loop effects are large primarily due to $ln\frac{m_z^2}{m_\mu^2} \simeq 13.5$ enhancements. Leading log 3 loop contributions turn out to be tiny (consistent with zero) as a result of an interesting cancellation. Overall, one finds in total[27]

$$a_\mu^{EW} = 154(1)(2) \times 10^{-11}, \tag{24}$$

where the first error comes from 2 loop hadronic triangle diagrams while the second is due to the Higgs mass uncertainty.

Combining eqs. (18),(23) and (24) gives the full Standard Model prediction

$$a_\mu^{SM} = 116591857(107) \times 10^{-11} \tag{25}$$

The central value will likely shift somewhat and the uncertainty should be reduced (perhaps by a factor of 2) as the 4 loop QED revisions are finalized and more important, as e^+e^- and τ data become consistent, so they can be legitimately averaged.

Comparing eqs. (15) and (25) gives (for now)

$$a_\mu^{exp} - a_\mu^{SM} = 173 \pm 134 \times 10^{-11} \tag{26}$$

That 1.3 sigma difference is currently not significant. It will be interesting to see how that difference evolves as a_μ^{exp} is improved and theoretical hadronic loop issues are clarified. It seems likely that an overall deviation from Standard Model expectations may result, but its significance may be limited to 2 or 3 sigma. That expectation is based on a comparison of other precision measurements such as $\alpha, G_\mu, m_z, sin^2\theta_w, m_w$ etc. with one another.[28] Such a comparison also depends on hadronic vacuum polarization effects and it favors the smaller $a_\mu^{Hadronic}(vac.pol.)$ in eq. (19) over (20), although such an analysis is suggestive rather than definitive.

16. Supersymmetry

If supersymmetry is correct, then all known elementary particles have partners with a different spin (by $\frac{1}{2}$ unit). Of course, supersymmetry must be broken such that the as yet undiscovered partners have large masses (which for simplicity are collectively labeled as m_{SUSY} here). Computing the one loop corrections to a_μ due to smuons, gauginos and sneutrinos along with the leading 2 loop effects, one finds the generic result[29]

$$\Delta a_\mu^{SUSY} \simeq \pm 130 \times 10^{-11} (\frac{100\ GeV}{m_{SUSY}})^2 \tan\beta \tag{27}$$

where the $\pm$ depends on the sign of the so-called μ term in the supersymmetry Lagrangian which mixes the two Higgs scalar doublets and $\tan\beta = \frac{<\varphi_2>}{<\varphi_1>}$. Since one expects $\tan\beta > 3$ and $m_{SUSY} \simeq 100 \sim 400\ GeV$, the contribution to a_μ from supersymmetry can be quite large. Indeed, it could easily exceed a_μ^{EW}. So, if SUSY particles exist in the multi 100 GeV region, they should be starting to appear in $a_\mu^{exp} - a_\mu^{SM}$. The only question is whether that quantity will have sufficient sensitivity to reveal it. Of course, if SUSY partners exist at the multi-100 GeV level, they will be discovered at the LHC. Knowing their masses and mixing with one another, one can perhaps then use a_μ to determine $\tan\beta$. However those events unfold, it is clear that a_μ^{exp} should be pushed as far as possible and a_μ^{SM} must attain comparable precision.

17. Other "New Physics" Effects

In addition to supersymmetry, there are many potential "New Physics" effects that could contribute significantly to a_μ. One generic possibility corresponds to models in which the bare mass $m_\mu^\circ = 0$ and the observed muon mass is generated by quantum loops.[30] (Examples are dynamical symmetry breaking, extra dimensions, softly broken SUSY, etc.) In such models similar quantum loops will induce a non-vanishing contribution to Δa_μ. For mass generation by scalar or gauge bosons of mass $\wedge$, one finds

$$\Delta a_\mu \simeq \frac{m_\mu^2}{\wedge^2} \tag{28}$$

Note that its coefficient is of $O(1)$ rather than $\frac{\alpha}{\pi}$. The current constraint in eq.(24) probes $\wedge \sim 2\ TeV$.

18. Outlook

During the 75 intervening years since Dirac explained why $g = 2$, anomalous magnetic moment experiments have been pushed to incredible precision. Currently,

$$\begin{aligned} g_e^{exp} &= 2.0023193043765(61) \\ g_\mu^{exp} &= 2.0023318406(16). \end{aligned} \tag{29}$$

Quantum corrections consistent with Standard Model expectations have clearly been discerned. They are now being probed at the few parts per trillion for g_e and a few parts per billion in the case of g_μ. At that level, effects of new particles and interactions (eg. supersymmetry) could start to be revealed. Unveiling their presence will, however, require further improved experimental precision and comparable heroic theory computations. Those pursuits continue to march forward. Along their path, new experimental techniques will be invented and technology will be pushed to its limits. In keeping up with those improvements, theory will be forced to similarly develop new computational techniques and confront strong interaction effects. Such challenges are key to the advancement of science. Experimental and theoretical physics move forward hand in hand for the pursuit of knowledge and the more we learn, the more inquisitive we become. The Dirac equation and anomalous magnetic moments have played and should continue to play an important role in that advancement.

References

1. P.A.M. Dirac, *Proc. Roy. Soc.* **A117**, 610 (1928); **A118**, 351 (1928).
2. See for example, A. Zee, "Quantum Field Theory in a Nutshell" Princeton Univ Press 2003. S. Weinberg, The Quantum Theory of Fields, Cambridge Univ. Press 1995.
3. W. Pauli, Handbuch der Physik (Julius Springer, Berlin 1932-33); *Rev. Mod. Phys.* **13**, 203 (1941).

4. Particle Data Group, K. Hagiwara et. al. *Phys. Rev.* **D66**, 010001 (2002).
5. S. Ferrara and E. Remiddi, *Phys. Lett.* **53B**, 347 (1974).
6. J. Nagle, E. Nelson and I. Rabi, *Phys. Rev.* **71**, 914 (1947); J. Nagle, R. Julian and J. Zacharias, *Phys. Rev.* **72**, 971 (1947).
7. G. Breit, *Phys. Rev.* **72**, 984L (1947).
8. J. Schwinger, *Phys. Rev.* **73**, 416L (1948).
9. P.A.M. Dirac, "Lectures on Quantum Field Theory", Belfer Graduate School of Science Monograph Series, New York 1966.
10. See for example, W. Marciano, "Elementary Particle Theory" in BNL Accel School (1983), ed M. Month.
11. G. Gabrielse and J. Tan, in "Cavity Quantum Electrodynamics", ed. P. Berman (San Diego: Academic) p267.
12. T. Kinoshita and M. Nio, *Phys. Rev. Lett.* **90**, 021803 (2003).
13. A. Czarnecki and W. Marciano, *Nucl. Phys. Proc.* **E76**, 245 (1999).
14. P. Mohr and B. Taylor, *Rev. Mod. Phys.* **72**, 351 (2000).
15. A. Czarnecki and W. Marciano, *Phys. Rev.* **D64**, 013014 (2001).
16. T. Kinoshita, in "The Gregory Breit Centennial Symposium (World Scientific 2001), eds. V. Hughes, F. Iachello and D. Kusnizov.
17. E821 G. Bennett et. al., *Phys. Rev. Lett.* **89**, 101804 (2002).
18. B. Lee Roberts, in "High Intensity Muon Sources", eds. Y. Kuno and T. Yokoi, World Scientific (1999) p69.
19. T. Kinoshita, talk at Lepton Moments Symposium, Cape Cod, Mass (2003).
20. T. Blum, HEP-LAT/0212018 (2002).
21. M. Davier, S. Eidelman, A. Hocker, and Z. Zhang, *Eur. Phys. J.* **C27**, 497 (2003).
22. M. Knecht and A. Nyffeler, *Phys. Rev.* **D65**, 073034 (2002).
23. J. Franzini, Lepton Moments Symposium, Cape Cod 2003.
24. R. Jackiw and S. Weinberg, *Phys. Rev.* **D5**, 2396 (1972); G. Altarelli, N. Cabibbo and L. Maiani, *Phys. Lett.* **B40**, 415 (1972); I. Bars and M. Yoshimura, *Phys. Rev.* **D6**, 374 (1972).
25. T. Kukhto et. al., *Phys. Rev.* **B371**, 567 (1992); A. Czarnecki, B. Krause and W. Marciano, *Phys. Rev.* **D52**, R2619 (1995); S. Peris, M. Perrottet and E. de Rafael, *Phys. Lett.* **B355**, 523 (1995); G. Degrassi and G. F. Giudice, *Phys. Rev.* **D58**, 053007 (1998).
26. A. Czarnecki, B. Krause and W. Marciano, *Phys. Rev. Lett.* **76**, 3267 (1996).
27. A. Czarnecki, W. Marciano and A. Vainshtein, *Phys. Rev.* **D67**, 073006 (2003).
28. W. Marciano, *Nucl. Phys. B (Proc. Suppl.)* **116**, 437 (2003).
29. T. Moroi, *Phys. Rev.* **D53**, 6565 (1996), Erratum **D56**, 4424 (1997); T. Ibrahim and P. Nath, *Phys. Rev.* **D57**, 478 (1998); D. Kosower, L. Krauss and N. Sakai, *Phys. Lett.* **B133**, 305 (1983).
30. W. Marciano, in "Radiative Corrections Status and Outlook", edited by B.F.L. Ward (World Scientific, Singapore, 1995) p403.

Pierre Ramond

Proceedings of the Dirac Centennial Symposium
Howard Baer and Alexander Belyaev

Dirac's Footsteps and Supersymmetry

Pierre Ramond
Institute for Fundamental Theory
Physics Department,
University of Florida, Gainesville, FL 32611, USA
E-mail: ramond@phys.ufl.edu

I am not interested in proofs.
I am only interested on how Nature works
P.A.M. Dirac

One hundred years after its creator's birth, the Dirac equation stands as the cornerstone of XXth Century physics. But it is much more, as it carries the seeds of supersymmetry. Dirac also invented the light-cone, or "front form" dynamics, which plays a crucial role in string theory and in elucidating the finiteness of $N = 4$ Yang-Mills theory. The light-cone structure of eleven-dimensional supergravity ($N = 8$ supergravity in four dimensions) suggests a group-theoretical interpretation of its divergences. We speculate they could be compensated by an infinite number of triplets of massless higher spin fields, each obeying a Dirac-like equation associated with the coset $F_4/SO(9)$. The divergences are proportional to the trace over a non-compact structure containing the compact form of F_4. Its nature is still unknown, but it could show the way to M-theory.

19. Dirac's equation

Much of modern physics starts with the Dirac equation. Its deceptively simple form

$$\gamma^\alpha p_\alpha \Psi = \not{p}\, \Psi = 0 , \tag{1}$$

obscured the significance of the simple algebraic relation for its anticommutator

$$\{\not p\,, \not p\} \;=\; p^2 \,. \tag{2}$$

Today we understand this as the seed of supersymmetry, which could be called protosupersymmetry. In the context of string models, this relation turned out to be very potent. There, the momentum was generalized to a momentum density along the string

$$P^\alpha(\sigma) \;=\; p^\alpha + \text{oscillators}\,, \tag{3}$$

suggesting that the momentum operator is the average of the momentum density

$$p^\alpha \;=\; < P^\alpha > \,. \tag{4}$$

If we assume that the Dirac matrices are themselves like dynamical variables (as envisioned by the Master himself[1])

$$\Gamma^\alpha(\sigma) \;=\; \gamma^\alpha + \text{oscillators}\,, \tag{5}$$

with

$$\gamma^\alpha \;=\; < \Gamma^\alpha > \,, \tag{6}$$

it is not unnatural[2] to expect that

$$\gamma^\alpha\, p_\alpha \;=\; < \Gamma^\alpha >< P_\alpha > \;\to\; < \Gamma^\alpha\, P_\alpha > \,, \tag{7}$$

which generalizes the Dirac equation to the string. If we write

$$\Gamma^\alpha(\sigma)\, P_\alpha(\sigma) \;=\; \sum_n F_n\, e^{in\sigma}\,, \tag{8}$$

we find the superVirasoro algebra

$$\begin{aligned} \{\, F_n\,, F_m \,\} &= 2\, L_{n+m}\,, \\ [\, L_n\,, F_m \,] &= (2n-m)\, F_{n+m}\,, \\ [\, L_n\,, L_m \,] &= (n-m)\, L_{n+m}\,. \end{aligned} \tag{9}$$

We should have read another of Dirac's papers[3] on the importance of $c-$numbers, which were found later by the late Joe Weiss. The rest is history as this algebra opened the Pandora box of theories with bosons and fermions, which we call supersymmetric.

20. Divergences and Group Theory

It is well-known how the running of the couplings in local field theory is related to the divergent part of some diagrams. For example, the one-loop beta function is given by

$$\beta = \frac{11}{3}\, I^{(2)}_{adj} - \frac{2}{3}\, I^{(2)}_{f} - \frac{1}{3}\, I^{(2)}_{H}\,, \tag{10}$$

where the $I^{(2)}_{adj,\,f,\,H}$ are the quadratic Dynkin indices associated with the adjoint (for the gauge bosons), with spin one-half Weyl fermions, and with complex spin zero fields, respectively. These "external" group theoretical factors are given by

$$\mathrm{Tr}\,(T^A_r\, T^B_r\,) \;=\; I^{(2)}_r\, \delta^{AB}\,, \tag{11}$$

where A, B run over the gauge group, and T^A_r are the representation matrices in the r representation. The other numerical factors stem from group theory of the "internal" space. As shown by Hughes,[4] they can be understood for massless particles as,

$$\frac{1}{3}\,(1 - 12h^2)\,, \tag{12}$$

where h is the helicity of the particle circulating around the loop. In some sense the square of the helicity can be viewed as the quadratic Dynkin index of the light-cone[5] spin group, although in four dimensions, it is only $SO(2)$, which is not much of a group.

Curtright[6] generalized this notion when he considered loop integrals coming from theories in higher dimensions, where the spin light-cone little group is more substantial. For instance, the one loop vacuum polarization in $N = 4$ Yang-Mills can be obtained directly in ten dimensions where the little group is $SO(8)$. Evaluate the loop integrals in four dimensions. The divergent part is proportional to

$$(-1)^f\left(\frac{1}{3}\, I^{(0)}\,(\delta^{(4)}_{\mu\nu}\, q^2 - q_\mu\, q_\nu) - \frac{4}{r}\, I^{(2)}\,(\delta^{(D)}_{\mu\nu}\, q^2 - q_\mu\, q_\nu)\right), \tag{13}$$

where $I^{(0)}$ is the dimension of the *transverse little group* representation, r its rank, and $I^{(2)}$ is the quadratic Dynkin index of the same. In the case of $N = 4$ Yang Mills, we have $D = 10$ and the transverse little group is $SO(8)$. With its triality property, the group theoretical factors are the same for bosons and fermions. This is the genesis of the cancellation of ultraviolet divergences for that theory. It is the peculiar properties of the transverse little group that leads to ultraviolet finiteness,[7] together with supersymmetry in the form of equality of fermions and bosons. This is true for *all* higher order Dynkin indices as well: since $SO(8)$ has rank four, it has three more independent Dynkin indices, of order $4, 6$, and 8, which are the same for bosons and fermions by triality. No other group has that property.

This is puzzling since all string theories stem from eleven-dimensional M-theory. The little group there is $SO(9)$, a totally unremarkable group, or so it seems. In eleven dimensions, supergravity is described by three fields,[8] the graviton $h_{\mu\nu}$, a Rarita-Schwinger fermion ψ_μ and a three-form boson $A_{\mu\nu\rho}$. Their physical degrees of freedom fall in three $SO(9)$ representations whose group-theoretical properties are summarized in the following table:

Field irrep	Ψ_i (1001)	h_{ij} (2000)	A_{ijk} (0010)
$I^{(0)}$	128	44	84
$I^{(2)}$	256	88	168
$I^{(4)}$	640	232	408
$I^{(6)}$	1792	712	1080
$I^{(8)}$	5248	2440	3000

The Dynkin indices have the remarkable property that they cancel between fermions and bosons

$$I^{(k)}_{\mathbf{128}} - I^{(k)}_{\mathbf{44}} - I^{(k)}_{\mathbf{84}} = 0 \; , \quad k = 0, 2, 4, 6 \; , \tag{14}$$

except for the highest invariant

$$I^{(8)}_{\mathbf{128}} - I^{(8)}_{\mathbf{44}} - I^{(8)}_{\mathbf{84}} = -192 \; . \tag{15}$$

This led Curtright to speculate that the theory is divergent because of this inequality. The lowest order divergent diagram that contains the eighth-order invariant is a three-loop four-graviton amplitude. While it is hopeless to calculate such a beast, this is the diagram for which there appears to be no local counterterm.[9]

21. Euler Triplets

Sometime ago, it was found[10] that this pattern of group-theoretical partial cancellations among three representations generalized to other $SO(9)$ representations.

There are three equivalent embeddings of $SO(9)$ inside the exceptional group F_4, much like the I- U- and V-spins for the embedding of $SU(2)\times U(1)$ inside $SU(3)$. As a result, one can associate with each F_4 representation *three* $SO(9)$ representations, whose properties are summarized in the character formula[11]

$$V_\lambda \otimes S^+ - V_\lambda \otimes S^- = \sum_c \mathrm{sgn}(c)\, U_{c\bullet\lambda} . \tag{16}$$

On the left-hand side, V_λ is a representation of F_4 written in terms of its $SO(9)$ subgroup, $S^\pm$ are the two spinor representations of $SO(16)$ written in terms of its anomalously embedded subgroup $SO(9)$, and $\otimes$ denotes the normal Kronecker product of representations. On the right-hand side, the sum is over c, the three elements of the Weyl group which map the Weyl chamber of F_4 into the (three times larger) chamber of $SO(9)$. Finally $U_{c\bullet\lambda}$ denotes the $SO(9)$ representation with highest Dynkin weight $c\bullet\lambda$, where

$$c\bullet\lambda = c(\lambda+\rho_{F_4}) - \rho_{SO(9)} ,$$

and ρ's are the sum of the fundamental weights for each group, and sgn(c) is the index of c. Thus to each F_4 representation corresponds a triplet, called Euler triplet. The three representations of supergravity appear in the trivial case associated with the singlet of F_4. Since

$$SO(16)\supset SO(9)\ , \qquad S^+ \sim \mathbf{128} = \mathbf{128}\ , \qquad S^- \sim \mathbf{128'} = \mathbf{44} + \mathbf{84}\ ,$$

the character formula reduces to

$$\mathbf{128} - \mathbf{44} - \mathbf{84} = \mathbf{128} - \mathbf{44} - \mathbf{84} .$$

In general, the representations describe (in light-cone variables) fields with spin greater than two. For each F_4 representation with Dynkin labels $[\,a_1\, a_2\, a_3\, a_4\,]$ one obtains three $SO(9)$ representations listed in order of increasing dimensions:

$$(2+a_2+a_3+a_4, a_1, a_2, a_3), (a_2, a_1, 1+a_2+a_3, a_4), (1+a_2+a_3, a_1, a_2, 1+a_3+a_4)$$

For spinor representations, the fourth entry is an odd integer. Euler triplets for which the largest representation is the spinor have equal number of fermions and bosons; this occurs whenever both a_3 and a_4 are even integers or zero.

22. Kostant Equation

We find here again the long hand of Dirac, for the minus sign in the character formula suggests that it is the index formula for a Dirac-like operator. This is Kostant's operator[12] associated with the coset $F_4/SO(9)$. The Clifford algebra over this coset

$$\{\Gamma^a\,,\Gamma^b\} \;=\; 2\,\delta^{ab}\;,\quad a,b=1,2,\ldots,16\;, \tag{17}$$

is generated by (256×256) matrices, and the Kostant equation is defined as

$$\not{K}\Psi \;=\; \sum_{a=1}^{16}\Gamma^a\,T^a\,\Psi \;=\; 0\;, \tag{18}$$

where T_a are the F_4 generators not in $SO(9)$, with commutation relations

$$[T^a\,,T^b\,] \;=\; i\,f^{\,ab\,ij}\,T^{ij}\;. \tag{19}$$

These are conveniently expressed in terms of copies of 26 oscillators with the usual Bose-like commutation relations[13]: $A_0^{[s]}$, $A_i^{[s]}$, $i=1,\cdots,9$, $B_a^{[s]}$, $a=1,\cdots,16$, and their hermitian conjugates, and where $s=1,2,3$. Under $SO(9)$, the $A_i^{[s]}$ transform as **9**, $B_a^{[s]}$ transform as **16**, and $A_0^{[s]}$ is a scalar. Note that the $B_a^{[s]}$ satisfy Bose-like commutation relations, even though they are $SO(9)$ spinors. The F_4 generators are then

$$T_{ij} = -i\sum_{s=1}^{4}\left\{\left(A_i^{[s]\dagger}A_j^{[s]} - A_j^{[s]\dagger]}A_i^{[s]}\right) + \frac{1}{2}\,B^{[s]\dagger}\,\gamma_{ij}\,B^{[s]}\right\}\;,$$

$$T_a = -\frac{i}{2}\sum_{s=1}^{4}\left\{(\gamma_i)^{ab}\left(A_i^{[s]\dagger}B_b^{[s]} - B_b^{[s]\dagger}A_i^{[s]}\right) - \sqrt{3}\left(B_a^{[s]\dagger}A_0^{[s]} - A_0^{[s]\dagger}B_a^{[s]}\right)\right\}\;.$$

One can just as easily have used the coordinate representation of the oscillators by introducing real coordinates u_i which transform as transverse space vectors, u_0 as scalars, and ζ_a as the space spinors. It is amusing to note that the internal cordinates span three exceptional Jordan algebras, which have been the subject of much interest as possible charge spaces.

The solutions of Kostant's equation are then simply described by a chiral superfield.[14] Listing only its highest weight components, it is of the form

$$\Phi_a = \theta^1\theta^8\, h(y^-, \boldsymbol{x},\, u_i,\, \zeta_a) \;+\; \theta^1\theta^4\theta^8\, \psi(y^-, \boldsymbol{x},\, u_i,\, \zeta_a)$$
$$+\,\theta^1\theta^4\theta^5\theta^8\, A(y^-, \boldsymbol{x},\, u_i,\, \zeta_a)\,. \tag{20}$$

The three fields are polynomials in three sets of "internal" bosonic coordinates. A possible violation of the spin-statistics connection is avoided when the fields are even functions of the ζ_a. This happens whenever a_3 and a_4 are even, but this is the case where each Euler triplet contains as many fermions as bosons. Hence there is an intriguing relation in this solution space between spin-statistics and equality between bosons and fermions. Supergravity is the trivial solution for which the three fields h, A, and ψ are independent of the internal coordinates.

23. A Zero Sum Game

Since the Dynkin indices of the product of two representations satisfy the composition law

$$I^{(n)}[\lambda \otimes \mu] \;=\; d_\lambda\, I^{(n)}[\mu] + d_\mu\, I^{(n)}[\lambda]\,,$$

where d is the dimension, it follows that the deficit in $I^{(8)}$ is always proportional to

$$d_\lambda(I^{(8)}_{S^+} - I^{(8)}_{S^-})\,,$$

where d_λ is the dimension of the F_4 representation that generates it. It is always of the same sign, so cancellation, if it happens will come only after summing over an infinite number of triplets.

Each triplet contains particles with spin greater than two. As we have shown,[14] they cannot have mass since they do not assemble in massive little group multiplets. The only evasion route from the well documented difficulties[15] with higher spin massless fields interacting with gravity theories is to allow for an infinite number, but to this date no such theory has been put forth, perhaps with good reason.

If indeed the divergences of supergravity all stem from the deficit in the eighth order Dynkin, we are led to the bizarre equation

$$\Delta\, I^{(8)} \;=\; -\,192 \sum_{a_1,a_2,a_3,a_4} d^{F_4}(a_1,a_2,a_3,a_4) \;=\; 0\,, \tag{21}$$

where $d^{F_4}(a_1, a_2, a_3, a_4)$ is the dimension of the F_4 representation with Dynkin labels $[\, a_1\, a_2\, a_3\, a_4\,]$. It can only vanish if the sum is over an infinite number of representations. In order to make sense of this at least two obstacles have to be surmounted. One is to specify the regularization procedure, and the second is to determine the subset of F_4 representations over which to take the sum.

We seek an algebraic structure which contains an infinite number of F_4 representations such that the (regulated) trace, or character, over the dimensions of these representations is zero. It would be wonderful if the triality properties of affine $E_6^{(1)}$ somehow became important. In quantum groups, zero traces can occur when the q parameter is a root of unity[a], so we are perhaps dealing with such an object. At any rate, it will be a miracle if such a structure exists, but then we are looking for a unique theory where miracles are expected on a daily basis.

Dimensions of F_4 representations are expressed as 24th order polynomials in their Dynkin indices. A zero-sum game can be clearly set up: assuming ζ-function regularization, or a quantum group, find a subset of F_4 representations with zero total dimension. The existence of such a set would be strong indication for a finite structure underlying eleven-dimensional supergravity.

Acknowledgments

I wish to thank Professor H. Baer for his kind hospitality and giving this mere mortal the honor to speak at the commemoration of one of the Gods of Physics. This work was supported in part by the US Department of Energy under grant DE-FG02-97ER41029

References

1. P. A. M. Dirac, *Proc. Royal Soc. (London)* **A117** 610: 1928
2. P. Ramond, *Phys. Rev.* **D3** 2415: 1971
3. P.A.M. Dirac, *Rev. Mod. Phys.* **34** 592: 1962
4. Richard J. Hughes, *Phys.Lett.* B97:246,1980
5. P.A.M. Dirac, *Rev. Mod. Phys.***21** 392:1948
6. T. Curtright, *Phys. Rev. Lett.* **48**, 1704(1982)
7. Lars Brink, Olof Lindgren, Bengt E.W. Nilsson. *Phys.Lett.* **B123**, 323(1983); Stanley Mandelstam, *Nucl.Phys.* **B213**149(1983)

[a] I thank E. Mukhin and P. di Francesco for an instructive discussion

8. E. Cremmer, B. Julia, J. Scherk , *Phys. Lett* **B76**, 409(1978)
9. S. Deser and J. H. Kay, *Phys. Lett.* **76B**, 400(1978); P. van Nieuwhenhuizen, *Phys. Rep.* **68**, 189(1981)
10. T. Pengpan and P. Ramond, *Phys. Rep.* **315**. 137(1999)
11. B. Gross, B. Kostant, P. Ramond, and S. Sternberg, *Proc. Natl. Acad. Scien.*, 8441 (1998)
12. B. Kostant, *Duke J. of Mathematics* **100**, 447(1999)
13. T. Fulton, *J. Phys. A:Math. Gen.* **18**, 2863(1985)
14. Lars Brink, Pierre Ramond, Xiao-zhen Xiong, JHEP 0210:058,2002; Hep-Th 0207253
15. M. A. Vasiliev, hep-th/0104246, and references therein

Stanley Deser

Proceedings of the Dirac Centennial Symposium
Howard Baer and Alexander Belyaev

P.A.M. Dirac and the Development of Modern General Relativity

Stanley Deser
Brandeis University
Department of Physics
Waltham, MA 02454, USA
E-mail: deser@brandeis.edu

I provide a very brief sketch of Dirac's Hamiltonian formulation of Einstein's theory, in its relation both to his own earlier work and to contemporary developments

It is a great honor for any theoretical physicist to speak at a memorial for Professor Dirac (like many others, I cannot bring myself to call him Paul!), whom I last saw while giving a Colloquium here in Tallahassee shortly before he died, when he was already quite frail. He went through the canonical behavior of sleeping during my talk, then awakening at the end with a perfectly reasonable question. He had expressed a desire to see me afterward in his office; there, he first asked me what was new in physics. I told him that he probably wouldn't be pleased to hear that there was a new, finite, quantum field theory, whereupon he asked – unhappily – what it was. When I told him it was N=4 supersymmetric Yang–Mills, he denied having heard of any of those words. After a suitable translation, his reaction was simply that either there was an error in the calculations or the model was really non-interacting; the latter opinion was then actually held by some pros as well. It was clear that his mind was made up about the bankrupcy of QFT, but then again, he was also one of the inventors of extended objects.

It is *de rigueur* to include a Dirac story in any lecture of which he is the subject, so that by now there are very few unknown ones. My favorite, because physically fraught, comes from Abdus Salam's introduction to Dirac's lecture at the famous 1968 Trieste Conference. When Dirac first came to St. John's, there was a traditional Christmas event at which the Maths tutors would pose a riddle to the incoming students – and here's what he drew. Three exhausted fishermen wash up on the beach with their haul, but are too sleepy to divide it up. At dawn, the first fisherman decides to take his 1/3 share and go home; this he does and throws the one leftover fish back into the sea. Number two, later, unaware of the first, acts identically, as does, finally the third one. What is the minimum size N of the original pile? Dirac's lightning-fast solution: N=–2! Not only does it foreshadow the Dirac sea but displays a fixed point invariance besides – every fisherman sees the *same* pile! [To keep my audience from calculating, here is the equation: $4N = 9P + 10$ where P is the pile faced by the late sleeper. Of course, $N = 25$ is the "correct", and so much duller, answer.]

What is *not* needed from me is an encomium: Dirac was a true Martian, a Hungarian(-in-law) one at that. I have seen the awe with which – for example – Feynman and Schwinger, neither otherwise overly impressionable – treated him, and he was right near Einstein and Bohr on Landau's famous logarithmic rankings. Instead, in the short time available, I will give a brief appreciation of Dirac's work in General Relativity (GR), which I was in a particular position (indeed a very competitive one!) to observe. [Incidentally, there exists[1] a very useful "Dalitz plot", referencing Dirac's complete works.] Here, I only cite the original papers, but not later lectures or reprises.

After its rapid initial successes, GR was very much a stepchild of theoretical physics research for many decades, until the early fifties. Even then the renaissance of which Dirac's work was part was primarily amongst the "natives", disconnected from the (then indifferent or hostile) field and particle theory mainstream. Indeed, the separation was traditional. Although GR was understood quickly after its discovery, neither Bohr nor Heisenberg, for example, ever ventured there, although the former did use the equivalence principle against Einstein in a famous debate on quantum mechanics at the 1927 Solvay Conference, and the latter was the first, in the late thirties, to understand why perturbative quantization of theories, with positive dimensional (self-)coupling constant would fail. Pauli of course started life with a text on GR, but despite continued interest, he never really contributed to it at the "Pauli" level. In later years, Schrödinger did venture into the field with some brilliant pedagogical expositions, but alas mostly into the morass of "unification by nonsymmetric metric" that occupied Einstein's own late years. Born explicitly wrote that once he understood GR, he vowed never to work on it. Thus (apart perhaps from Jordan and Klein), Dirac was unique among the creators of quantum mechanics to work seriously on GR. The first, and completely isolated, attempts at treating GR as a dynamical system were probably those of the early thirties by Rosenfeld,[2] and by Bronstein[3] in the USSR. Tragically, the latter name coincided with Trotsky's real one and he disappeared early in the Stalin purges.

Looking at Dirac's own earlier work, one is struck first by two seemingly disconnected, but in fact indicative, themes. The first, physics in deSitter (dS) space,[4] comes (as usual) out of nowhere. One of the results here, that "masslessness" means a non-vanishing mass parameter for spin 1/2 has had profound repercussions, via Lee and Gursey,[5] on supergravity.[6,7] [Amusingly, Dirac worked only in dS rather than in AdS, and so required this mass to be imaginary, $m \sim \sqrt{-\Lambda}$, instead of real as in the natural SUGRA domain, AdS!] [More formally, Dirac exploited the Weitzenbock identity for gravity, exactly as he first did in his Dirac equation paper to discover the $g = 2$ factor for the electron in a background magnetic field. To analyze propagation, one must first square the Dirac equation to get a wave operator. Now when the partial derivative is replaced by a covariant one, the square is more complicated: apart from factors of 1/2, i etc., one has schematically

$$\not{D}^2 \equiv \frac{1}{2}\{\gamma,\gamma\}DD + \frac{1}{4}[\gamma,\gamma][D,D] \tag{1}$$

the last term vanishes for ordinary derivatives, but the commutator $[D, D]$ just defines a curvature (a "field strength") and in the gravitational case, it all reduces to the scalar curvature, which is of course proportional to the cosmological constant Λ. Consequently,

$$(\not{D} + m)(\not{D} - m) = D^2 - (\Lambda + m^2) \tag{2}$$

for a spinor. For all higher (also integer) spins, things get even more interesting, as described in[8].] The other thread is of course Dirac's abiding interest in Lagrangian/Hamiltonian dynamics, locality and "unusual" systems. Two of the great[9,10] articles on formulations of dynamical systems are probably the most cited, but there is one other (rather different) significant example, introduction[11] of what Dirac calls "homogenous coordinates". This is nothing but a variant of the Jacobi form of the traditional action principle in flat space in which time and Hamiltonian form a new, "n + first", conjugate (q, p) pair, so that in terms of the extended set, and of the Lagrange multiplier λ that keeps the "true number" of variables n, the original Lagrangian

$$L = \sum_{i=1}^{n} p_i \dot{q}_i - H(p,q) \tag{3a}$$

becomes

$$L = \sum_{i=1}^{n+1} p_i \dot{q}_i + \lambda(p_{n+1} + H(p,q)) \,. \tag{3b}$$

where $\overline{H} = 0$ has a root $p_{n+1} = -H(p,q)$. Instead, in our independent (ADM) work,[12] understanding that the Einstein–Hilbert (or any other diffeo-invariant) action was necessarily an "already parametrized" system *a la* (3b) was an essential, beautiful, confirmation of the Hamiltonian formulation. There is no *a priori* passage back from (3b) to (3a), in contrast to Jacobi: Instead one fixed a gauge, $q_{n+1} = t$ and solved the constraint for its conjugate p_{n+1} quite arbitrarily. Surprisingly, Dirac never explicitly invoked the Jacobi formulation for GR; his true work in GR was concentrated in three papers,[13,14,15] and I am sure one of his motivations was to apply his general formalisms to this challenging system. [For an idea of the changing research directions that characterized the era, the Proceedings[16] of several conferences bracketing this time are instructive: the first, in Bern, was in 1955, then came Chapel Hill in 1957, Royaumont in 1959 and Warsaw in 1962.] The density of competing groups in the field was so low that the dilute approximation can be used in assessing Dirac's contribution.[a]

Let us then summarize the second[14] of Dirac's contributions, to which[13] was the prelude; it contains his major results and shows his reasoning. The paper starts with the fundamental kinematical recognition that a 3+1 decomposition of the field variables, rather than maintaining full covariance, is essential to any Hamiltonian – and hence quantization – development. Using projections with respect to a t=const. surface, and noting that the $g_{\mu 0}$ are only lapses and shifts (as they were later called) and so not dynamical, he inferred that the time development of any dynamical (*i.e.*, not involving

[a] Our own (ADM) history, which is closest to Dirac's approach, began[12] in Spring 1958 with an analysis of the linearized approximation of GR about flat space (massless spin 2 field), totally unaware of Dirac, let alone Rosenfeld, Bronstein or Pauli and Fierz. It was only after we had worked out the main framework of the full theory that we heard Dirac's talk on his Hamiltonian procedure in Royaumont. It turned out that we were extremely close in both spirit and formalism, though there were also rather different aspects as well. But this talk is about Dirac.

$g_{\mu 0}$) gravitational variable η is of the form

$$\dot{\eta} = \int d^3x[N\xi_L + N_i\xi^i] \,, \quad N \equiv (-g^{00})^{-1/2} \,, \quad N_i \equiv g_{0i} \tag{4}$$

and hence must be derived from a Hamiltonian

$$H = \int d^3x(NH_L + N_iH^i) \,; \quad H_L \equiv {}^3\!R + (p_{ij}^2 - p_{ij}^2/2) \,, \quad H^i = -D_jp^{ij} \tag{5}$$

where H_L, H^i depend as shown on the conjugate (but *not* independent) pairs of spatial tensors (g_{ij}, p^{ij}), which ("weakly") obey the standard set of Poisson brackets. Here p^{ij} are essentially the second fundamental forms. The (H_L, H^i) are four ("weak" or "secondary") constraints on the (g_i, p^{ij}), and they are understood to be essentially the G^0_μ components of the Einstein tensor, because those do not depend on second time derivatives, easily noted from the Bianchi identities, $\partial_0 G^0_\mu \sim (\partial_i + \Gamma)G$: the right side has only second time derivatives, so the left, G^0_μ, must only have first derivatives. [The above picture, although reached by different reasoning, also holds in the Lagrangian approach.] At this point, the natural – but hardly unique – coordinate choice $g_{\mu 0} = -\delta_{\mu 0}$ is invoked to yield what he calls H_{MAIN}, which is just $\int d^3x\mathcal{H}_L$, after dropping the total 3-divergence that is the linear part of the 3-scalar curvature, *i.e.*, the (ADM) energy. [Dirac's third work[15] is also concerned with establishing the form of the field's energy; I do not have the space to discuss it here.] The paper goes on to discuss general issues of constrained Hamiltonian dynamics, in the spirit of,[9,10] in order to understand coordinate fixing and reduction of the apparent number of variable pairs by the constraints. The choice of time is the obvious minimal surface one, $p^i_i = 0$, leaving 5 pairs and finally a spatial harmonic gauge is invoked for the remainder. The resulting P.B. are discussed in a perturbative way. The paper ends in a short section entitled "Quantization", which simply says that a complete set of commuting variables can be chosen as the unimodular, divergenceless part of the spatial metric, together with matter variables; hence the wave function is unconstrained as it depends only on this reduced space, and thus simply obeys the Schrödinger equation. As the paper ends, Dirac is of course worrying about quantum problems: he notes that the t=const. surfaces must remain spacelike, which means a positive signature for g_{ij} – *i.e.*, positive determinant. Since this quantity is also related to the "energy density", he states that violation could occur very near point sources, basically due to negative gravitational self-energy. The concluding sentence is: "The gravitational treatment of point particles thus brings in one further difficulty, in addition to the usual ones in the quantum theory." This is rather curious coda since the above problems are really already relevant classically, and of course they are very different from the perturbative nonrenormalizability issues that have dominated all subsequent studies. After this pioneering foray, Dirac's publications in the field waned. Perhaps the most significant later paper was[17] on conformally invariant extensions of GR.

After this brief historical sketch of Dirac's work, I will close by emphasizing that, in an era when geometry was (for real relativists) the vital guiding thread to the mysteries of GR, the Hamiltonian approach provided a rapprochement to ordinary gauge QFT, in which the field is a dynamical system with degrees of freedom, asymptotic boundary conditions and global conserved quantities correlated to the chosen asymptotics. That approach also serves well (as proved necessary much later) when the cosmological constant does not vanish and when the more intimate gravity-matter unifications of supergravity made their appearance. Although Dirac lived to see SUGRA as well as (pre-revolution) string theory, he was, like Moses, unable to enter this promised land, even though he had also been an important predecessor in so many ways. We all stand on Dirac's shoulders.

References

1. *The Collected Works of P.A.M. Dirac, 1924–1948*, R.H. Dalitz, ed. (Cambridge Univ. Press, 1995); there is, alas, no companion volume (yet?) for 1949–84.
2. L. Rosenfeld, Ann. d. Physik **5**, 113 (1930); Ann. IHP **2** 25 (1932).
3. M. Bronstein (exact reference not available).
4. P.A.M. Dirac, Ann. Math. **36**, 657 (1935).
5. F. Gursey and T.D. Lee, Proc. Nat. Acad. Sci. **49**, 179 (1963).
6. P.K. Townsend, Phys. Rev. **D15**, 2808 (1977).
7. S. Deser and B. Zumino, Phys. Rev. Lett. **38**, 1433 (1977).
8. S. Deser and A. Waldron, Phys. Lett. **B513**, 137 (2001) and references therein.
9. P.A.M. Dirac, "Forms of Relativistic Dynamics," Rev. Mod. Phys. **21**, 392 (1989).
10. P.A.M. Dirac, "Generalized Hamiltonian Dynamics," Can. J. Math. **2**, 129 (1950). 4.
11. P.A.M. Dirac, "Homogeneous Variables in Classical Dynamics," Proc. Camb. Phil. Soc. **29**, 389 (1933).
12. R. Arnowitt and S. Deser Phys. Rev. **113**, 745 (1959); R. Arnowitt, S. Deser and C.W. Misner *ibid* **116**, 1322 (1959). For a summary of our subsequent papers, see "The Dynamics of General Relativity" in *Recent Developments in Gravitation*, L. Witten, ed. (Wiley, NY 1962).

13. P.A.M. Dirac, "The Theory of Gravitation in Hamiltonian Form," Proc. R. Soc. Lond. **A246**, 333 (1958).
14. P.A.M. Dirac, "Fixation of Coordinates in the Hamiltonian Theory of Gravitation," Phys. Rev. **114**, 924 (1959).
15. P.A.M. Dirac, "Energy of the Gravitational Field," Phys. Rev. Lett. **2**, 368 (1959).
16. Helv. Phys. Acta Suppl. IV (1956) (Bern); *Les Theories Relativistes de la Gravitation*, Eds. du CNRS (1962) (Royaumont); Revs. Mod. Phys. **29** #3 (1957) (Chapel Hill); *Proc. of Theory of Gravitation*, Gauthier–Villars, Paris (1964) (Warsaw).
17. P.A.M. Dirac, Proc. Roy. Soc. **A333**, 403 (1973); S. Deser, Ann. Phys. **59**, 248 (1970).

Brian D. Serot

Proceedings of the Dirac Centennial Symposium
Howard Baer and Alexander Belyaev

Building Atomic Nuclei with the Dirac Equation

Brian D. Serot
Physics Department and Nuclear Theory Center
Indiana University
Bloomington, IN 47405
E-mail: serot@iucf.indiana.edu

The relevance of the Dirac equation for computations of nuclear structure is motivated and discussed. Quantitatively successful results for medium- and heavy-mass nuclei are described, and modern ideas of effective field theory and density functional theory are used to justify them.

24. Introduction

To understand how to build atomic nuclei with the Dirac equation, we will begin by asking some simple questions. What are the basic nuclear properties that we are trying to correlate and predict? Why use hadrons (rather than quarks and gluons) as the degrees of freedom? Why use the Dirac equation rather than the Schrödinger equation to describe the dynamics? How can we build a simple model of nuclear matter that reproduces the empirical equilibrium properties and that can be extended to calculations of medium- and heavy-mass nuclei? How does the Dirac approach predict the nuclear shell model? And how can we relate the hadronic description of nuclei to the underlying strong-interaction theory of quantum chromodynamics (QCD)?

The basic properties of nuclei provide stringent constraints on any nuclear theory. An accurate description of these properties is necessary for any useful predictions or extrapolations. We will concern ourselves primarily with bulk and single-particle nuclear properties, as listed below; a more detailed discussion can be found in Refs. [1,2].

We certainly want to reproduce the observed shapes of nuclei: the interior density of a heavy nucleus should be relatively constant, since the nuclear forces "saturate" at the equilibrium density of nuclear matter (roughly $\rho_0 \approx 0.15\,\text{fm}^{-3}$). Moreover, the nucleus should have a well-defined surface, with the density decreasing from 90% to 10% of its central value over a distance of roughly 2 fm. Finally, because of saturation, the radius R of a nucleus should scale according to $R \approx A^{1/3}\,1.1\,\text{fm}$, where $A = N + Z$ is the total number of neutrons (N) plus protons (Z).

The total energy of the nucleus should follow the "liquid drop" formula

$$E = -a_1 A + a_2 A^{2/3} + a_3 Z^2/A^{1/3} + a_4 (N-Z)^2/A + \cdots \,, \tag{1}$$

where typical values for the a_i coefficients are given in Ref. [1].

The particle spectrum is determined by the qualitative features of the single-particle potential. In a nonrelativistic (Schrödinger) language, the central potential is midway between a harmonic oscillator and a square well; this shape determines the ordering of the levels as a function of the orbital angular momentum. (See Ref. [1], Figs. 57.1 and 57.2.) In addition, the spin-orbit potential is strong, which is instrumental in determining the major shell closures and, hence, the shell model. We will see below how these features are reproduced in a description based on the Dirac equation.

These simple nuclear features are the ones we will focus on. We expect that they can be described adequately by a single-particle equation with an effective, one-body interaction. Such an approach has many names, depending on the system being studied and on the practitioner: "shell model", "mean-field theory", "Kohn–Sham" density functional theory, etc. Our goal is to correlate (fit) a modest number of nuclear bulk and single-particle data and then to predict other, similar data as well as possible.

25. Why Use Hadrons?

Well, why not? Our focus is on long-range nuclear characteristics, and all measured observables are colorless. (In fact, most of the observables relevant to us are dominated by the *isoscalar* part of the interaction.) Moreover, hadronic variables (baryons and mesons) are efficient, since hadrons are the particles that are observed in experiments. Colored quarks and gluons participate *only in intermediate states*, and such "off-shell behavior" is unobservable; by using hadrons, we expend no theoretical effort combining quarks and gluons into color singlets that can actually be observed.

So we pick the most efficient degrees of freedom by choosing hadrons. We will have to parametrize the nuclear hamiltonian anyway, since we cannot compute its true form from QCD, and hadronic variables, if combined in all forms consistent with the underlying symmetries, provide sufficient flexibility for our parametrization. We cannot guarantee that a single-particle hadronic approach will be successful in describing the observables of interest, but we want to see how well we can do.

26. Why Use the Dirac Equation?

To motivate the Dirac equation as straightforwardly as possible, compare the particle spectrum (and fine structure) in a light atom with the spectrum in a heavy nucleus. An example of the former is given in Ref. [3], while an early example of the latter is given in Ref. [4], which is reproduced in Fig. 57.3 of Ref. [1]. The most striking result is that it is impossible to draw the atomic fine structure to scale, since the splittings are roughly 1/10,000 as large as the major-level splittings (at least for the deeply bound atomic levels). In contrast, the nuclear spectrum shows that the "fine" structure is really "gross"; the fine-structure splittings are as large as the major-level splittings to within a factor of two!

The implication is that there *must* be some relativistic effects that are *important* in nuclei (unlike light atoms), and thus it is much more natural to use the Dirac equation to describe the quasi-particle nucleon wave functions. We will now try to understand this result by building a simple model of uniform nuclear matter.

27. A Simple Model of Nuclear Matter

We consider a model first proposed by Walecka,[5] which contains nucleons (ψ) and neutral (isoscalar) Lorentz scalar (φ) and vector (V^μ) mesons. This model is often referred to as "quantum hadrodynamics I" (or QHD–I, for short). The lagrangian density for this model (using the conventions of Ref. [6] and suppressing counterterms for simplicity) is

$$\mathcal{L} = \overline{\psi}(i\gamma_\mu\partial^\mu - M)\psi + \frac{1}{2}(\partial_\mu\varphi\,\partial^\mu\varphi - m_{\rm s}^2\varphi^2) - \frac{1}{4}(\partial_\mu V_\nu - \partial_\nu V_\mu)^2$$
$$+ \frac{1}{2}m_{\rm v}^2 V_\mu V^\mu + g_{\rm s}\overline{\psi}\psi\varphi - g_{\rm v}V^\mu\,\overline{\psi}\gamma_\mu\psi \,. \tag{2}$$

The included degrees of freedom are the minimal ones that will allow us to understand the qualitative features of the nuclear many-body system, which is our goal. We will describe the system in terms of Dirac quasi-particles moving in classical meson mean fields, an approximation that we will elaborate on and justify later. Note that the baryon current (density) $\overline{\psi}\gamma_\mu\psi$ is conserved.

It is important to emphasize that the Lorentz scalar and vector fields are *effective fields* that are introduced to parametrize the nucleon–nucleon (NN) interaction. The quanta of these fields never appear "on the mass shell" as real particles in any of the calculations discussed here. They are analogous to the phonons that describe electron–electron interactions inside a metal.

If one computes the NN interaction using one-boson exchange [purely for illustration, since the coupling constants g_s and g_v in Eq. (2) are large], one finds a short-range repulsion (from V^μ) and a mid-range attraction (from φ), which is characteristic of the NN force. Explicit pion exchange is of minor importance for our observables of interest; isoscalar, scalar and vector fields are dominant for the bulk and single-particle properties of heavy nuclei (and some *multi-pion* exchange is simulated by our effective fields anyway).

Consider nuclear or neutron matter at zero temperature. We can treat the mesons at the mean-field level by taking

$$\varphi \longrightarrow \langle\varphi\rangle \equiv \varphi_0 \;, \quad V^\mu \longrightarrow \langle V^\mu\rangle \equiv V_0\,\delta^{\mu 0} \;, \tag{3}$$

where $\langle\mathbf{V}\rangle \equiv 0$, since we assume that we are in the rest frame of the uniform matter. Note that φ_0 and V_0 are *constants.*

Why should mean meson fields yield a reasonable description of the system, since QHD–I is a strong-coupling theory? Our goal is to construct an approximate energy functional of the scalar (ρ_s) and baryon (ρ_B) densities and to fit the parameters in this functional to bulk nuclear properties. The mean meson fields give us a convenient way to do this, since they satisfy the *mean-field equations*:

$$\varphi_0 = \frac{g_\mathrm{s}}{m_\mathrm{s}^2}\,\rho_\mathrm{s} \equiv \frac{g_\mathrm{s}}{m_\mathrm{s}^2}\sum_i^{\mathrm{occ}} \overline{\psi}_i\psi_i \;, \tag{4}$$

$$V_0 = \frac{g_\mathrm{v}}{m_\mathrm{v}^2}\,\rho_\mathrm{B} \equiv \frac{g_\mathrm{v}}{m_\mathrm{v}^2}\sum_i^{\mathrm{occ}} \psi_i^\dagger\psi_i \;, \tag{5}$$

$$[i\gamma_\mu\partial^\mu - g_\mathrm{v}\gamma^0 V^0 - \underbrace{(M - g_\mathrm{s}\varphi_0)}_{M^*}]\,\psi = 0 \;, \tag{6}$$

where "occ" signifies the occupied quasi-particle levels. (In infinite matter, we sum over states with both spin projections and with momentum $k \leq k_{\rm F}$, where $k_{\rm F}$ is the Fermi momentum.)

We solve these equations for stationary quasi-particle states; the problem is self-consistent, since $\rho_{\rm s} = \rho_{\rm s}(M^*)$ both determines and depends on the wave functions.[5,6] The nuclear/neutron matter energy function(al) then becomes

$$\mathcal{E}_{\rm MFT} = \frac{g_{\rm v}^2}{2m_{\rm v}^2}\rho_{\rm B}^2 + \frac{m_{\rm s}^2}{2g_{\rm s}^2}(M - M^*)^2 + \frac{\lambda}{\pi^2}\int_0^{k_{\rm F}} dt\, t^2\,(t^2 + M^{*2})^{1/2}\ , \tag{7}$$

where the baryon density is

$$\rho_{\rm B} \equiv \frac{\lambda k_{\rm F}^3}{3\pi^2}\ , \tag{8}$$

and the isospin degeneracy is $\lambda = 2$ for symmetric ($N = Z$) nuclear matter and $\lambda = 1$ for pure neutron matter ($Z = 0$). (Note that, by definition, the Coulomb force between protons is turned *off*.)

One can now minimize the energy density $\mathcal{E}$ with respect to $\rho_{\rm B}$ to find the equilibrium point, and use the *empirical* equilibrium point of nuclear matter (density $= \rho_0 \approx 0.15\,{\rm fm}^{-3}$, binding energy $= e_0 \approx 16\,{\rm MeV}$) to determine the two unknown ratios

$$\frac{g_{\rm s}^2}{m_{\rm s}^2} \quad \text{and} \quad \frac{g_{\rm v}^2}{m_{\rm v}^2}\ ,$$

which are expressed more conventionally (and less dimensionally) as

$$C_{\rm s}^2 \equiv \frac{g_{\rm s}^2 M^2}{m_{\rm s}^2} = 357.4 \quad \text{and} \quad C_{\rm v}^2 \equiv \frac{g_{\rm v}^2 M^2}{m_{\rm v}^2} = 273.8\ . \tag{9}$$

The resulting nuclear/neutron matter binding curves and the self-consistent effective mass M^* as functions of the density are shown in Figs. 1 and 2 of Ref. [6]. The important features of these results are:

- Symmetric nuclear matter is a self-bound liquid with an equilibrium point as defined above. This illustrates the "saturation" of nuclear forces.
- Pure neutron matter is (generally) unbound at all densities. This reflects the positive symmetry-energy coefficient [a_4 in Eq. (1)] that enters when the number of neutrons and protons is different.[a]

[a]In QHD–I, this coefficient is too small. One must add a ρ meson, which couples to the difference of proton and neutron densities, to achieve an accurate result. See Ref. [6].

- The nucleon effective mass at equilibrium density is roughly $M_0^* \approx 0.6M$. This shows that the scalar mean field is roughly -400 MeV at equilibrium; the corresponding vector mean field is roughly 300 MeV, and the two fields cancel to produce the relatively small binding energy of 16 MeV. We turn now to a discussion of this point.

What causes the nuclear matter saturation and the relatively small binding energy? Let's expand $\mathcal{E}/\rho_{\text{B}}$ from Eq. (7) in powers of k_{F}:[6]

$$\begin{aligned}\mathcal{E}_{\text{MFT}}/\rho_{\text{B}} = M &+ \left[\frac{3k_{\text{F}}^2}{10M} - \frac{3k_{\text{F}}^4}{56M^3} + \frac{k_{\text{F}}^6}{48M^5} - \cdots\right] + \frac{g_{\text{v}}^2}{2m_{\text{v}}^2}\rho_{\text{B}} - \frac{g_{\text{s}}^2}{2m_{\text{s}}^2}\rho_{\text{B}} \\ &+ \frac{g_{\text{s}}^2}{m_{\text{s}}^2}\frac{\rho_{\text{B}}}{M}\left[\frac{3k_{\text{F}}^2}{10M} - \frac{36k_{\text{F}}^4}{175M^3} + \cdots\right] + \left(\frac{g_{\text{s}}^2\rho_{\text{B}}}{m_{\text{s}}^2 M}\right)^2\left[\frac{3k_{\text{F}}^2}{10M} - \cdots\right] \\ &+ \left(\frac{g_{\text{s}}^2\rho_{\text{B}}}{m_{\text{s}}^2 M}\right)^3\left[\frac{3k_{\text{F}}^2}{10M} - \cdots\right] + \cdots \end{aligned} \tag{10}$$

The lowest-order Lorentz scalar and vector contributions (which are proportional to ρ_{B}) *set the scale* for the large mean fields. [See Eqs. (4) and (5).] This scale is consistent with chiral QCD counting rules,[7,8] but these two terms *cancel almost exactly* in the binding energy, leading to an anomalously small remainder. However, they *add constructively* in the spin-orbit interaction, leading to appropriately large spin-orbit splittings in nuclei.[9,10,11]

It is important to notice the *different behavior* of the vector and scalar interaction terms in Eq. (10). Whereas the vector interaction enters at only linear order in ρ_{B}, the scalar interaction enters at all orders; moreover, the leading scalar term at every order in ρ_{B} looks exactly the same, and they all add constructively. These terms are precisely what one gets by shifting the nucleon mass in the nonrelativistic kinetic energy term $3k_{\text{F}}^2/10M$ from $M \to M^* \approx M - g_{\text{s}}^2\rho_{\text{B}}/m_{\text{s}}^2$. These additional, repulsive, velocity-dependent interactions reduce the strength of the lowest-order, attractive scalar contribution and are crucial for establishing the location of the equilibrium point of nuclear matter. Thus the different behavior of the vector and scalar interactions leads to *large relativistic interaction effects* in the nuclear matter energy density. In contrast, the relativistic corrections to the kinetic energy (the nonleading terms in the first pair of square brackets) are indeed small; this is *not* where the important "relativity" is.

28. Mean-Field Theory for Nuclei

We now compute the bulk and single-particle properties of atomic nuclei using essentially the same simple lagrangian discussed above. Our treatment follows that of Ref. [10], which is more than twenty years old, but which is still sufficient to illustrate the important points. We will discuss modifications and more modern treatments later.

The basic idea is to allow the mean meson fields to be spatially dependent, and we will consider only spherically symmetric nuclei for simplicity. We again look for stationary quasi-nucleon states, and so the mean-field equations become[6]

$$\nabla^2\varphi_0(r) - m_s^2\varphi_0(r) = -g_s\rho_s(r) \ , \tag{11}$$

$$\nabla^2 V_0(r) - m_v^2 V_0(r) = -g_v\rho_B(r) \ , \tag{12}$$

$$\{-i\boldsymbol{\alpha}\cdot\nabla + g_v V_0(r) + \beta[M - g_s\varphi_0(r)]\}\,\psi_\alpha(\mathbf{x}) = \varepsilon_\alpha\psi_\alpha(\mathbf{x}) \ . \tag{13}$$

These are coupled, nonlinear, differential equations that must be solved *self-consistently.* They are sometimes called Dirac–Hartree equations[12,6] but are more accurately described as Kohn–Sham equations,[13] as we discuss in more detail below.

As one might expect, an accurate description of nuclear properties is not possible using only nucleons and isoscalar mesons. One must extend the model to include at least the Coulomb interaction between protons and an isovector ρ meson that allows for a more realistic description of the nuclear symmetry energy. (See Refs. [10,12] for details.) The augmented model now contains four adjustable parameters:

$$g_s \ , \quad g_v \ , \quad g_\rho \ , \quad m_s \ .$$

(The heavy meson masses are fixed at some "large" mass scale that is roughly equal to the nucleon mass M.) The couplings are fitted to the equilibrium point of symmetric nuclear matter and to the nuclear matter symmetry energy; the length scale, which is determined by m_s, is set by fixing this parameter to reproduce the rms charge radius of a doubly magic nucleus, such as ^{40}Ca.

Many nuclear structure calculations have been carried out within this relativistic mean-field theory (RMFT) framework. (See, for example, Refs. [10,12] or the extensive list of references in Ref. [6].) One finds that the bulk properties of nuclei are well reproduced even in this relatively simple mean-field theory. Moreover, the single-particle spectrum reveals the well-known nuclear shell structure; this comes *for free*, since the parameters are fitted to the *bulk* properties of nuclear matter (and one nuclear length scale).

Extensions of this simple model have been made to "fine-tune" the results. In the 1980's, numerous authors added terms involving nonlinear interactions of the scalar field:

$$\mathcal{L}' = -\frac{1}{3!}\kappa\varphi^3 - \frac{1}{4!}\lambda\varphi^4 , \tag{14}$$

and in the 1990's, various practitioners added vector self-couplings, like

$$\mathcal{L}'' = +\frac{1}{4!}\zeta g_{\rm v}^4(V_\mu V^\mu)^2 , \tag{15}$$

as well as other nonlinear and gradient-coupling terms, some motivated by the ideas of effective field theory; see the discussion below. (Many calculations in these extended models are cited in Ref. [6]. For an alternative approach that uses only nucleons in a lagrangian that contains numerous powers of fermion fields, see Ref. [14] and references therein.)

These additional nonlinearities can be interpreted in terms of many-body nuclear forces, and they introduce additional density dependence into the nuclear energy functional, which allows it to more accurately reproduce the true energy functional. The new parameters are fitted either to additional nuclear matter properties, or to other theoretical calculations of nuclear matter (based on the Schrödinger equation), or to a selected set of data from finite nuclei. The basic conclusion from these extended calculations is that the successful qualitative features predicted by the original simple models persist, but the quantitative accuracy increased by nearly two orders of magnitude over a period of twenty years. For a comparison of the accuracy of results obtained with different collections of parameters, see, for example, Refs. [7,8]. For some recent state-of-the-art *predictions* of this approach (that is, calculations of nuclei that are not included in the fitting procedures), see Ref. [15].

But it still remains for us to understand at a deeper level why these simple relativistic mean-field calculations can do such an excellent job of reproducing certain nuclear observables. For this, we must study ...

29. Modern Developments

The discussion in this section is a synopsis of the formalism presented in Refs. [6,7,13,17], which is based on the ideas of modern effective field theory (EFT) and density functional theory (DFT). The interpretation of the earlier, successful results using EFT/DFT puts them on a firm theoretical basis.

First of all, we interpret QHD as a nonrenormalizable EFT. This means that it contains *known long-range interactions* that are constrained by the underlying QCD symmetries, plus a complete (but non-redundant) set of *generic short-range interactions*, i.e., "contact" and "gradient" terms. The borderline between short and long ranges is characterized by the *breakdown scale* Λ of the EFT; empirically, we find that $\Lambda \approx 600\,\text{MeV}$ for QHD.[8]

If we ignore strangeness, then only nucleons and pions are "real" (stable) particles. The other field quanta are always virtual and just let us parametrize the NN interaction. As in any lagrangian theory, there are different ways to choose the generalized coordinates (fields), but some coordinates may be more efficient than others.[8,14]

The QHD EFT lagrangian explicitly exhibits the symmetries of QCD: The global, chiral $SU(2)_L \times SU(2)_R$ symmetry is nonlinear, approximate, and spontaneously broken.[6] The remaining global, isovector subgroup $SU(2)_V$ is realized linearly. It is straightforward (but usually tedious) to include electromagnetic interactions through the familiar local U(1) gauge symmetry.[7]

The basic strategy for using the QHD lagrangian has been developed over the last several years.[16,7,6] First, assign an index ν to each term in the lagrangian:

$$\nu = d + n/2 + b\,, \tag{16}$$

where d is the number of derivatives (not counting those that act on nucleon fields),[b] n is the number of nucleon fields, and b is the number of non-Goldstone bosons.

Now organize the lagrangian in powers of ν and truncate. This gives an expansion in inverse powers of a heavy mass scale $\Lambda \approx M$, which has been shown to be reliable in calculations of medium- and heavy-mass nuclei.[7,8] Practically speaking, in the nuclear many-body problem, this expansion is

[b] Time derivatives acting on nucleon fields will generally bring down factors of the nucleon mass or energy, which are not small compared to Λ.

in powers of k_F/M, where k_F is the Fermi momentum at equilibrium nuclear density ($k_F/M \approx 1/3$).

Use the truncated lagrangian to construct an energy functional, which is to be interpreted within the DFT framework: We approximate the functional using *factorized* densities or fields, which produces a mean-field form of the functional. Expand it as a power series in density and momentum (by counting powers of ν) and fit the remaining parameters to a restricted set of experimental data.[8] These data typically include nuclear binding energies, prominent features of the nuclear electromagnetic charge form factors, and single-particle energy splittings for the least-bound orbitals.[7] Define a set of Kohn–Sham (KS) single-particle orbitals that satisfy differential equations obtained by extremizing the energy functional with respect to the densities and fields. This procedure guarantees that all of the source terms in these equations are *local.* The KS orbitals are tailored to the generation of the ground-state density, and they include short-range and correlation effects adequately, if the mean-field energy functional is a good approximation to the true energy functional.[17]

The mean-field energy functional constructed above omits some long-range contributions, which are generally nonlocal and nonanalytic functions of the densities. These contributions can be added systematically, by computing loop integrals using the well-known rules of EFT.[18] The effect of these loop contributions on the energy functional of atomic nuclei is an important topic for future study.

30. Summary

The most important points in the preceding discussion can be summarized as follows:

- The Dirac equation provides an economical and natural way to describe bulk nuclear properties and the nucleon single-particle spectrum, with the correct spin-orbit force (that is, the nuclear shell model) arising *automatically.*
- Kinematical relativistic effects are small in nuclei, but dynamical relativistic effects from the interactions are important.
- Modern QHD EFT's incorporate the basic symmetries of QCD.
- The mean-field approach to heavy nuclei is really DFT, implemented through KS quasi-particle orbitals. The tested validity and accuracy of our truncation procedure for both fitted *and predicted*

results shows that we really know something about the energy functional for cold nuclear matter near equilibrium density!
- The energy functional can be extended beyond the mean-field parametrization using well-defined rules of EFT to compute the long-range contributions of loop integrals. This has been done recently.[18]
- The QHD/EFT/DFT/KS formalism provides a *true representation* of QCD in the low-energy nuclear domain.

The basic message of this talk is: the Dirac equation is relevant for nuclear-structure physics, even though you might not expect it to be. Our quantitative successes justify its usage, but the modern theoretical ideas of EFT and DFT explain why it works.

Acknowledgments

I am grateful to the organizers of the Dirac Centennial Symposium for the opportunity to visit Tallahassee and to share my ideas about the structure of atomic nuclei. I am also pleased to acknowledge long, friendly, and fruitful collaborations with Dick Furnstahl, Ying Hu, Hua-Bin Tang, and Dirk Walecka during the course of these studies. This work was supported in part by the US Department of Energy under contract no. DE–FG02–87ER40365.

References

1. A. L. Fetter and J. D. Walecka, Quantum Theory of Many-Particle Systems (McGraw–Hill, N.Y., 1971), chs. 11 and 15.
2. J. D. Walecka, Theoretical Nuclear and Subnuclear Physics (Oxford U. Press, N.Y., 1995), part I.
3. J. D. Bjorken and S. D. Drell, Relativistic Quantum Mechanics (McGraw–Hill, N.Y., 1964, 1998), fig. 4–2.
4. M. G. Mayer and J. H. D. Jensen, Elementary Theory of Nuclear Shell Structure (Wiley, N.Y., 1955), p. 58.
5. J. D. Walecka, *Ann. of Phys.* **83**, 491 (1974).
6. B. D. Serot and J. D. Walecka, *Int. J. Mod. Phys.* E **6**, 515 (1997).
7. R. J. Furnstahl, B. D. Serot, and H.-B. Tang, *Nucl. Phys.* **A615**, 441 (1997); **A640**, 505(E) (1998).
8. R. J. Furnstahl and B. D. Serot, *Nucl. Phys.* **A671**, 447 (2000).
9. W. H. Furry, *Phys. Rev.* **50**, 784 (1936).
10. C. J. Horowitz and B. D. Serot, *Nucl. Phys.* **A368**, 503 (1981).
11. R. J. Furnstahl, J. J. Rusnak, and B. D. Serot, *Nucl Phys.* **A632**, 607 (1998).
12. B. D. Serot and J. D. Walecka, *Adv. Nucl. Phys.* **16**, 1 (1986).

13. R. J. Furnstahl and B. D. Serot, *Comments Nucl. Part. Phys.* **2**, A23 (2000).
14. J. J. Rusnak and R. J. Furnstahl, *Nucl. Phys.* **A627**, 495 (1997).
15. M. A. Huertas, *Phys. Rev.* C **66**, 024318 (2002).
16. R. J. Furnstahl, B. D. Serot, and H.-B. Tang, *Nucl. Phys.* **A598**, 539 (1996).
17. W. Kohn, *Rev. Mod. Phys.* **71**, 1253 (1999).
18. Y. Hu, Ph.D. thesis, Indiana University (2000), unpublished.

Vernon Barger

Proceedings of the Dirac Centennial Symposium
Howard Baer and Alexander Belyaev

New Focus on Neutrinos

Vernon Barger

Physics Department, University of Wisconsin, Madison, WI 35706

Dirac spent the spring of 1929 at the University of Wisconsin. During his visit he was interviewed by a colorful reporter, Roundy, from the Wisconsin State Journal newspaper. The interview was published in the April 30, 1929 edition, under the header: ROUNDY INTERVIEWS PROFESSOR DIRAC: AN ENJOYABLE TIME WAS HAD BY ALL. The following excerpts from the interview reveal much about Dirac's personality.

> The other afternoon I knocks at the door of Dr. Dirac's office in Sterling Hall and a pleasant voice says "Come in." And I want to say here and now that this sentence "come in" was about the longest one emitted by the doctor during our interview. He sure is all for efficiency in conversation. It suits me. I hate a talkative guy.
> "What do you like best about America?" says I.
> "Potatoes," says he.
> "Same here," says I. "Do you like to read the Sunday comics?"
> "Yes," says he, warming up a bit more than usual.
> "And now I want to ask you something more – do you ever run across a fellow that even you can't understand?"
> "Yes," says he.
> "This will make a great reading for the boys down at the office," says I. "Do you mind releasing to me who he is?"
> "Weyl," says he. ...he let loose a smile as we parted and I knew that all the time he had been talking to me he was solving some problem that no one else could touch.

The mention of Weyl brings me to the subject of this talk. The two component Weyl fermions are the minimal fermion degrees of freedom. For neutrinos, the relevant Weyl fermions are ν_L (under CPT, $\nu_L \leftrightarrow \nu_R^c$), the members of SU(2) doublets with charged leptons, and N_R, SU(2) singlets (under CPT, $N_R \leftrightarrow N_L^c$). A Dirac neutrino mass term $m_D\overline{\nu}_L N_R + h.c.$ has $\Delta L = 0$, where L is lepton number. Majorana mass terms are $\frac{1}{2}(M_R \overline{N}_L^c N_R + \text{h.c.}$ and $\frac{1}{2}(m_T \overline{\nu}_L \nu_R^c + \text{h.c.})$, which have $\Delta L = \pm 2$. The presence of mixed Majorana and Dirac masses with $M_R \gg m_D$ leads to the see-saw mechanism that provides a natural explanation of small m_ν, $m_{\nu_\tau} \sim m_D^2/M_R$, for $m_D \sim m_t$ and $M_R \sim \frac{1}{2} \times 10^{15}$ GeV. A consequence of the see-saw is that the light neutrinos are Majorana.

31. Model Building

Many models of neutrino masses and mixings have been proposed. Supersymmetric Grand Unified Theories invoke the see-saw mechanism, mass textures, and flavor symmetries at the unification scale. An alternative is low-energy new physics in which neutrino masses are generated as loops; an example is supersymmetry with R-parity violation, in which case neutrinos are also Majorana. New symmetries have been proposed for the neutrino sector such as A_4 and extra U(1) symmetries. Large extra dimensions with sterile neutrinos in the bulk and active neutrinos on the brane have also been considered.

A number of models are already excluded by the data. In particular those with solar oscillation solutions of small angle MSW, vacuum, and LOW types and those with large active-sterile mixings are eliminated, as well as schemes with large CPT violations. The early theoretical preference for the small angle MSW solar solution has engendered revisions of models. Since we presently have no general theory for the origin of masses, progress can only be made through model building and comparison of consequences with experimental observations.

32. Neutrino Counting

From studies of e^+e^- annihilation at the Z-resonance pole at the Large Electron Positron collider the invisible width of the Z boson has been determined. The experimental value $N_\nu = 2.984 \pm 0.008$ is close to the number expected from 3 active light neutrinos, though the value is 2σ low.

The cosmic microwave background (CMB) anisotropies and Big Bang Nucleosynthesis also probe the number of neutrinos. Even with the precision WMAP CMB data, the constraint on N_ν from the CMB is somewhat weak,

$N_\nu \lesssim 8$. However, the WMAP data accurately determine $\eta_{10} = 274\Omega_b h^2$, where Ω_b is the mass fraction of baryons in units of the critical density and h is the reduced Hubble constant (with value $h = 0.72 \pm 0.08$ found by the HST Key Project). With the WMAP determination of η_{10}, agreement of the BBN predictions of the abundances of primordial He^4 and deuterium gives stringent constraints on the number of relativistic neutrino degrees of freedom. The BBN upper-bound on N_ν is now 3.2 at 95%C.L., consistent with 3 neutrinos, giving no support to the possible existence of sterile neutrinos.

33. Neutrino Mixing

The dramatic increase in our knowledge of neutrino properties has come from observational evidence of neutrino oscillations. These neutrino flavor changes require that the neutrino flavor states, ν_α are not the same as the neutrino mass eigenstates, ν_i. The eigenstates are related by a unitary matrix V,

$$\nu_\alpha = \sum V^*_{ai} \nu_i \,. \tag{1}$$

For 3 neutrinos, the mixing matrix V is specified by three rotation angles $\theta_a, \theta_x, \theta_s$ and three CP-violating phases δ, φ_2 and φ_3. V can be conveniently written as the matrix product

$$V = \begin{bmatrix} 1 & 0 & 0 \\ 0 & c_a & s_a \\ 0 & -s_a & c_a \end{bmatrix} \begin{bmatrix} c_x & 0 & s_x e^{-i\delta} \\ 0 & 1 & 0 \\ -s_x e^{i\delta} & 0 & c_x \end{bmatrix} \begin{bmatrix} c_s & s_s & 0 \\ -s_s & c_s & 0 \\ 0 & 0 & 1 \end{bmatrix} \begin{bmatrix} 1 & 0 & 0 \\ 0 & e^{i(\frac{1}{2}\varphi_2)} & 0 \\ 0 & 0 & e^{i(\frac{1}{2}\varphi_3+\delta)} \end{bmatrix} \tag{2}$$

where c_i denotes $\cos\theta_i$ and s_i denotes $\sin\theta_i$. The angle θ_a, customarily denoted as θ_{23}, governs the oscillations of atmospheric neutrinos, the angle θ_s (θ_{12}) describes solar neutrino oscillations, and the angle θ_x (θ_{13}) is an unknown angle that is bounded by reactor neutrino experiments at short distances ($L \simeq 1$ km). The oscillation probabilities are independent of the Majorana phases φ_2 and φ_3. Vacuum neutrino oscillations are given by

$$P(\nu_\alpha \to \nu_\beta) = \left| \sum V_{\beta j} e^{-\frac{i m_j^2}{2E_\nu}} V^*_{\alpha i} \right|^2 , \tag{3}$$

where the m_j are the neutrino eigenmasses. The oscillation probabilities depend only on differences of mass-squared. The oscillation arguments for the atmospheric and solar phenomena are

$$\Delta_a \equiv \frac{\delta m_a^2 L}{4E_\nu}\,, \qquad \Delta_s \equiv \frac{\delta m_s^2 L}{4E}\,, \tag{4}$$

where

$$\delta m_a^2 = m_3^2 - m_1^2 \qquad \delta m_s^2 = m_2^2 - m_1^2\,. \tag{5}$$

In solar neutrino oscillations matter effects modify the vacuum oscillation probabilities. The scattering of ν_e on electrons changes the probability amplitude $\sin^2 2\theta_s$ in vacuum to

$$\sin^2 2\theta_s^m = \frac{\sin^2 2\theta_s}{\left(\frac{A}{\delta m_s^2} - \cos 2\theta_s\right)^2 + \sin^2 2\theta_s} \tag{6}$$

in matter. Here $A = 2\sqrt{2}\,G_F\,N_e\,E_\nu$ with N_e the electron density. The oscillation amplitude in matter is enhanced if $\delta m_s^2 > 0$.

34. Solar Neutrinos

Decades of study of neutrinos from the Sun have convincingly established that neutrino oscillations are the cause of the deficits of 1/3 to 1/2 in the measured electron-neutrino flux relative to the Standard Solar Model expectations. The water Cherenkov experiments of SuperKamiokande and SNO measure the high energy neutrinos ($E \gtrsim 5$ MeV) from the ^{8}B chain, the Chlorine experiment includes the intermediate energy neutrinos from ^{7}Be and pep, and the Gallex and Sage experiments have dominant contributions from the *pp* chain that powers the Sun. Until recently, the interpretation of the deficits depended on comparisons with SSM predictions of the flux. With the SNO experiment, which directly measures the total active neutrino flux via neutral currents, the test of the oscillation hypothesis becomes robust.

The SNO experiment utilizes a heavy water target and measures the following processes:

$$\text{Charged-Current (CC):}\quad \nu_e + d \to e^- + p + p \tag{7}$$

$$\text{Neutral-Current (CC):}\quad \nu_x + d \to \nu_x + n + p \tag{8}$$

$$\text{Elastic-Scattering (ES):}\quad \nu_x + e^- \to \nu_x + e^- \tag{9}$$

The CC/NC ratio establishes the oscillations of ν_e to ν_μ and ν_τ flavors,

$$\text{CC/NC} = \nu_e/(\nu_e + \nu_\mu + \nu_\tau)\,. \tag{10}$$

The charged-current signal is found to be suppressed by 5.3σ from the neutral-current signal. (Note: only ν_e are produced in the Sun; the ν_μ and ν_τ fluxes are a consequence of oscillations.)

The day and night energy spectra of charged-current events are potentially sensitive to matter effects on oscillations that occur when the neutrinos travel through the Earth. In global fits to neutrino data from all experiments, including the SNO and SuperKamiokande day/night spectra, regions of the solar oscillation parameters have been determined, as shown in Fig. 1. The Large Mixing Angle (LMA) solution is strongly preferred, with the LOW solution allowed only at 3σ C.L. The best fit to the solar data is $\delta m_s^2 = 5.6 \times 10^{-5}\,\text{eV}^2$ and $\tan^2\theta_s = 0.39$.

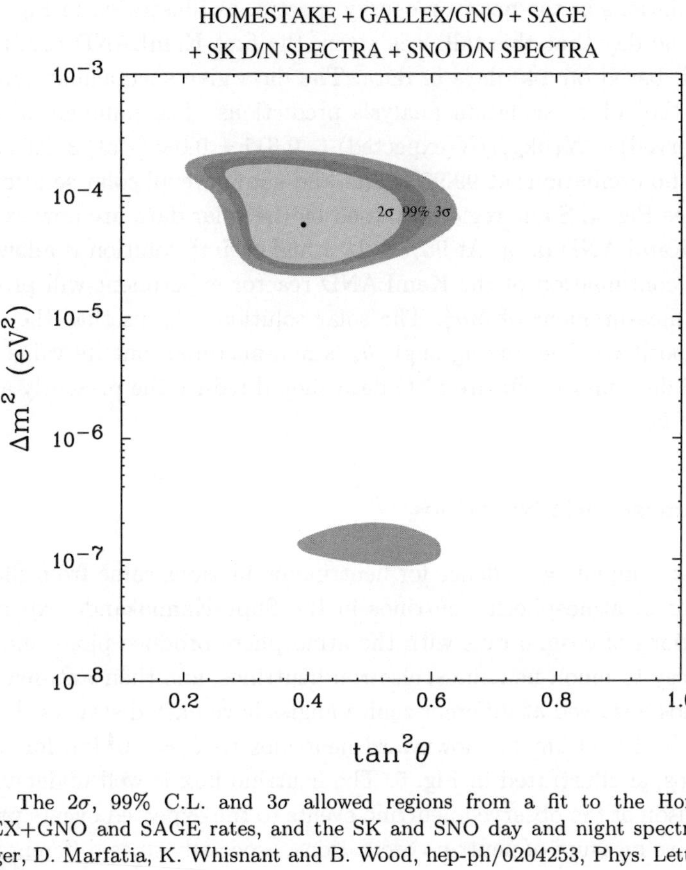

Fig. 1. The 2σ, 99% C.L. and 3σ allowed regions from a fit to the Homestake, GALLEX+GNO and SAGE rates, and the SK and SNO day and night spectra. From V. Barger, D. Marfatia, K. Whisnant and B. Wood, hep-ph/0204253, Phys. Lett. **B537**, 179 (2002).

35. Reactor Anti-neutrinos

The KamLAND experiment measures the electron anti-neutrino flux at the Kamiokande detector from surrounding reactors. The dominant reactor is at $L = 160$ km and the average distance from the sources is $L \sim 180$ km. The measured reaction is $\overline{\nu}_e + p \to e^+ + n$. If CPT invariance holds, which is expected in quantum field theory, then $P(\overline{\nu}_e \to \overline{\nu}_e) = P(\nu_e \to \nu_e)$. If the LMA solar solution is correct, then reactor anti-neutrinos should also disappear due to oscillations. For any other solar oscillation solution, no disappearance would be observed at KamLAND. The pre-KamLAND expectations for 3 years data assuming the LMA oscillation parameters are shown in Fig. 2 by the narrow ellipses superimposed on the present solar LMA region. With sufficient data, the KamLAND experiment should "see" the oscillations in the positron energy spectra, as illustrated in Fig. 3.

On the day that this talk was given, the first KamLAND results were released, based on 145 days of data. The data give spectacular confirmation of the solar oscillation analysis predictions. The numbers of events $(N(\text{observed}) - N(\text{bkg}))/N(\text{expected}) = 0.611 \pm 0.085(\text{stat}) \pm 0.041(\text{syst})$ exclude no oscillations at 99.95% C.L. and eliminate all solar solutions but LMA; see Fig. 4. Some regions allowed by the solar data are now excluded by the KamLAND data. At 95% C.L. a higher δm_s^2 solution is allowed.

The continuation of the KamLAND reactor experiment will provide a precise measurement of δm_s^2. The solar solution tells us that the sign of δm_s^2 is positive. The mixing angle θ_s is non-maximal, but its value is still not well determined. Future SNO data should reduce the presently allowed range of θ_s.

36. Atmospheric Neutrinos

The first compelling evidence for neutrino oscillations came from the measurement of atmospheric neutrinos in the SuperKamiokande experiment. Interactions of cosmic rays with the atmosphere produce pions and kaons that decay to muon-neutrinos, electron-neutrinos, and their anti-neutrinos. Neutrinos observed at different zenith angles have path distances that vary from $L \sim$ 10–30 km for downward neutrinos to $L \sim 10^4$ km for upward neutrinos, as illustrated in Fig. 5. The neutrino flux is well understood. A comparison of the observed neutrino events to the expected events provides a sensitive measure of neutrino oscillations, especially since different ranges of neutrino energies can be studied. It is concluded from the SuperK data

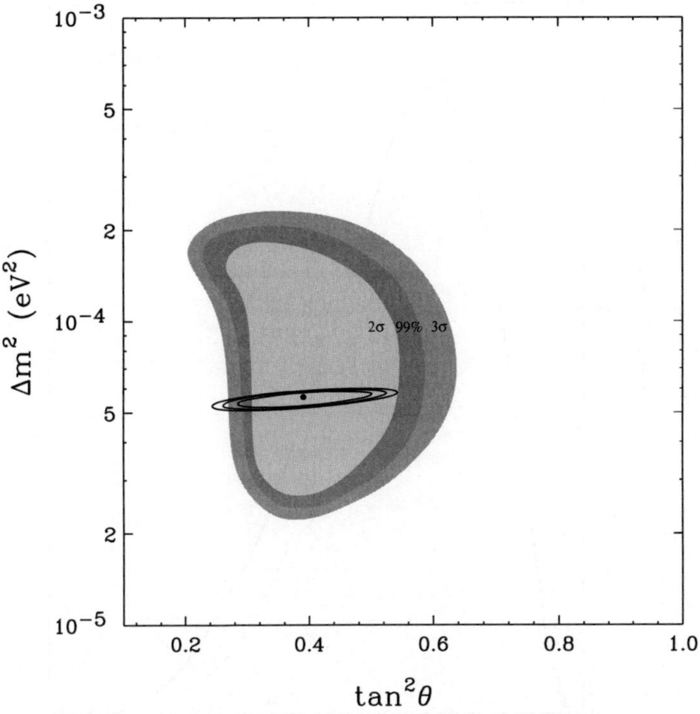

Fig. 2. Projection of how well KamLAND will determine the oscillation parameters with three years of data accumulation assuming an LMA solution. Data were simulated at the best-fit LMA parameters. The ellipses are the 2σ, 99% C.L. and 3σ KamLAND regions. From V. Barger, D. Marfatia, K. Whisnant and B. Wood, hep-ph/0204253, Phys. Lett. **B537**, 179 (2002).

that muon-neutrinos oscillate to tau-neutrinos with nearly maximal mixing, $\sin^2 2\theta_a > 0.92$ at 90% C.L. The best fit to the mass-squared difference of these oscillations is $\delta m_a^2 = 2.5 \times 10^{-3}\,\mathrm{eV}^2$, with a 90% C.L. range of $(1.6–3.9) \times 10^{-3}\,\mathrm{eV}^2$. No evidence for electron-neutrino oscillations was found, indicating that the mixing angle θ_x is small, consistent with the CHOOZ reactor limit.

37. Absolute Neutrino Mass

Neutrino oscillations tell us nothing about the absolute scale of neutrino masses. The standard technique for probing the absolute mass is to study the end-point region of the electron spectrum in tritium beta-decay. The effect of a non-zero neutrino mass is to suppress and cut off the electron

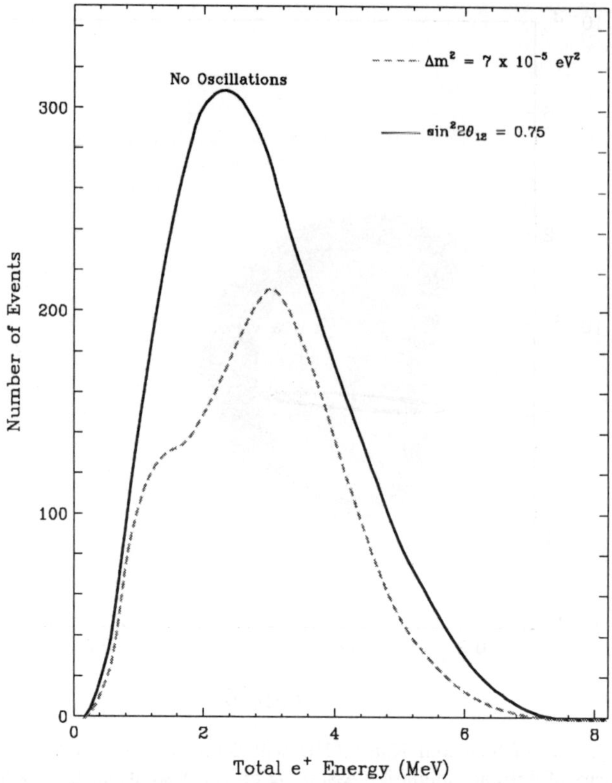

Fig. 3. Kamland's sensitivity to Δm^2 is unprecedented. In three years it will easily be able to discriminate between only slightly different values of Δm^2 in the LMA region. From V. Barger, D. Marfatia, and B. Wood, hep-ph/0011251, Phys. Lett. **B498**, 53 (2001).

distribution at the highest energies. The effective neutrino mass that could be determined in beta-decay is

$$m_\beta^2 = \sum |V_{ei}|^2 m_{\nu_i}^2 \,. \tag{11}$$

The present limit from the Troitsk and Mainz experiments is $m_\beta < 2.2$ eV. Future sensitivity down to $m_\beta = 0.35$ eV is expected in the KATRIN experiment, which will begin in 2006.

Absolute neutrino mass can also be probed in cosmology through the large scale structure of the Universe. The galaxy power spectrum is influenced by the sum of neutrino masses, even down to 0.1 eV. Neutrinos that are more massive cluster more on large scales. The analysis of the 2dF Galaxy Redshift Survey gives a limit of $\sum m_\nu < 2.2$ eV on the masses of

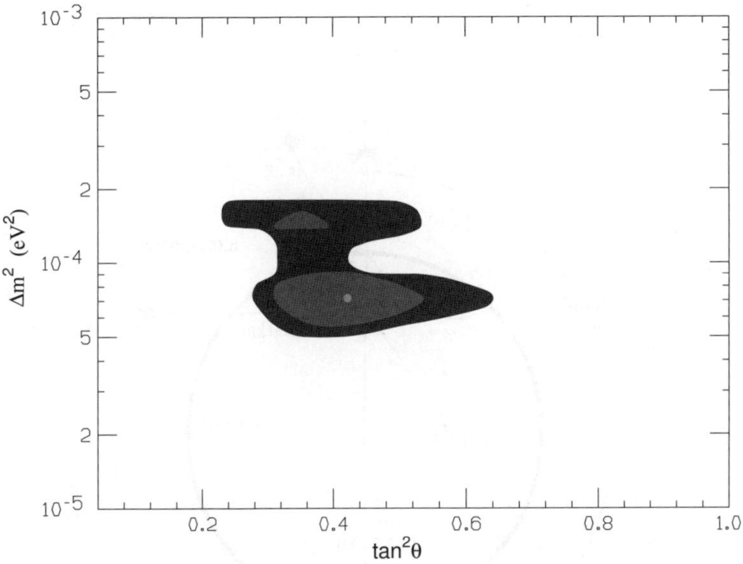

Fig. 4. The 2σ and 3σ allowed regions from a combined fit to KamLAND and solar neutrino data. The best-fit point is at $\Delta m^2 = 7.1 \times 10^{-5}$ eV^2 and $\tan^2\theta = 0.42$. From V. Barger and D. Marfatia, Phys. Lett. **B555**, 144 (2003).

degenerate neutrinos, or about 0.7 eV for each neutrino. An improved limit of $\sum m_\nu < 0.7$ eV was obtained by analysis of the WMAP and other CMB data in conjunction with the 2dFGRS and the Lyman alpha forest power spectrum. However, the important role of the Lyman alpha forest data in the latter limit makes this $\sum m_\nu$ constraint less conclusive due to questions about the uncertainties on the Lyman alpha forest data. Excluding the Lyman alpha forest data, the limit on the summed neutrino masses is 1 eV.

All neutrino masses are linked to the lightest mass by the values of δm_a^2 and δm_s^2 determined by the neutrino oscillation studies. Since the sign of δm_a^2 is unknown, there are two possible neutrino mass hierarchies, as illustrated in Fig. 6.

Another probe of absolute neutrino mass is neutrinoless double-beta decay ($0\nu\beta\beta$), provided that neutrinos are Majorana. The decay rate depends on the ν_e–ν_e element of the mixing matrix:

$$M_{ee} = \sum \left| V_{ei}^2 m_i \right| . \tag{12}$$

The prediction is insensitive to θ_x because it is small. Setting $\theta_x = 0$ and δm_s^2, and taking $m_1 < m_2 < m_3$, the following relations are obtained in

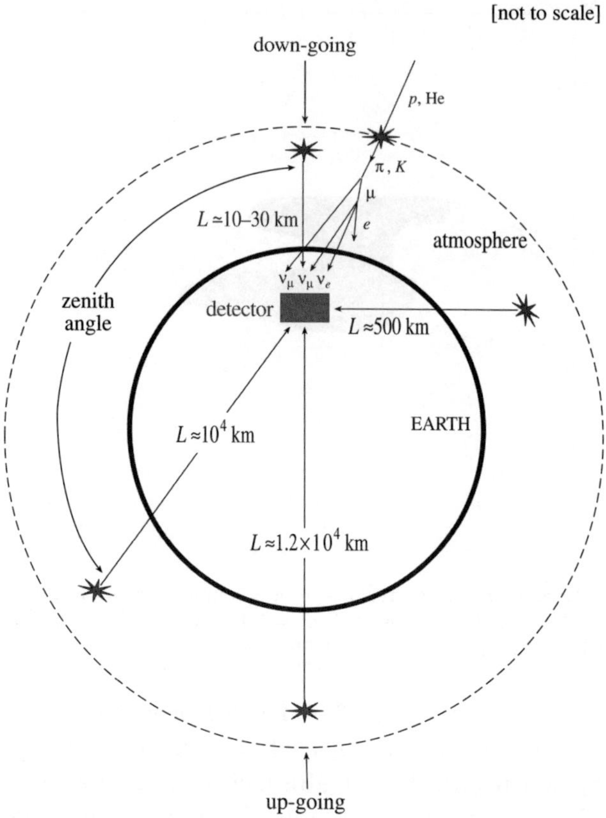

Fig. 5. A schematic view of the different zenith angles of atmospheric neutrinos and distances they travel before detection.

the two hierarchies:

$$\begin{array}{cc}
\underline{\text{normal mass hierarchy}} & \underline{\text{inverted mass hierarchy}} \\
m_1 = m_2 \equiv m_e & m_2 = m_3 \equiv m_e \\
m_3 = \sqrt{m_e^2 + \Delta} & m_1 = \sqrt{m_e^2 - \Delta} \\
\Sigma = 2m_e + \sqrt{m_e^2 + \Delta} & \Sigma = 2m_e + \sqrt{m_e^2 - \Delta} \\
M_{ee} = m_e \left| c_s^2 + s_s^2 e^{i\varphi_2} \right| & M_{ee} = m_e \left| c_s^2 + s_s^2 e^{i\varphi_2} \right|
\end{array} \tag{13}$$

where $\Delta = \left|\delta m_a^2\right|$. For a given measured value of M_{ee} both upper (since $\theta_s \neq \pi/4$) and lower bounds are implied for Σ. These bounds are displayed in Fig. 7. Thus, neutrinoless double-beta decay can constrain neutrino dark matter, whose relic density is given by

$$\Omega_\nu h^2 = \Sigma/(94\ \text{eV})\,. \tag{14}$$

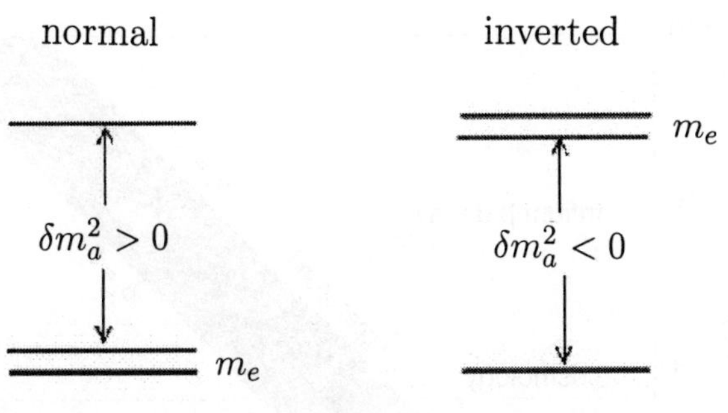

Fig. 6. The patterns of relative mass differences in normal (left) and inverted (right) neutrino mass hierarchies.

The present upper limit on M_{ee} is $M_{ee} < 0.46$ eV, with an overall factor of 3 uncertainty associated with nuclear matrix elements. A detection of neutrinoless double beta decay has been reported, but this experimental result is highly controversial.

38. Future Agenda

A summary of present knowledge of neutrino parameters is given in Table 1, along with the future projects that will improve this knowledge.

The near term agenda is to confirm atmospheric neutrino oscillations in accelerator experiments and improve the accuracy on those oscillation parameters. Experiments that measure ν_μ disappearance will establish the first oscillation minimum in $P(\nu_\mu \to \nu_\mu)$. The K2K experiment from KEK to SuperK, a distance of $L = 250$ km, is restarted following the reconstruction of the SuperK detector. The MINOS experiment from Fermilab to the Soudan mine, at a distance of $L = 750$ km, will begin in 2005. It is expected to obtain 10% precision on δm_a^2 and $\sin^2 2\theta_a$ in 3 years running. The CERN to Gran Sasso (CNGS) experiments, ICARUS and OPERA, at a distance $L = 730$ km, are expected to begin in 2007.

The appearance of ν_τ in $\nu_\mu \to \nu_\tau$ oscillations should be observed in the CNGS experiments.

The appearance of ν_e in $\nu_\mu \to \nu_e$ oscillations is the most critical measurement, since this depends on $\sin^2 2\theta_x$. By combining ICARUS/MINOS/OPERA data, it should be possible to establish whether $\sin^2 2\theta_x > 0.01$ at 95% C.L. Precision measurement of θ_x awaits future off-axis neutrino beam experiments proposed at Fermilab and the Japan Hadron Facility (JHF). Off-axis beams have nearly monoenergetic neutri-

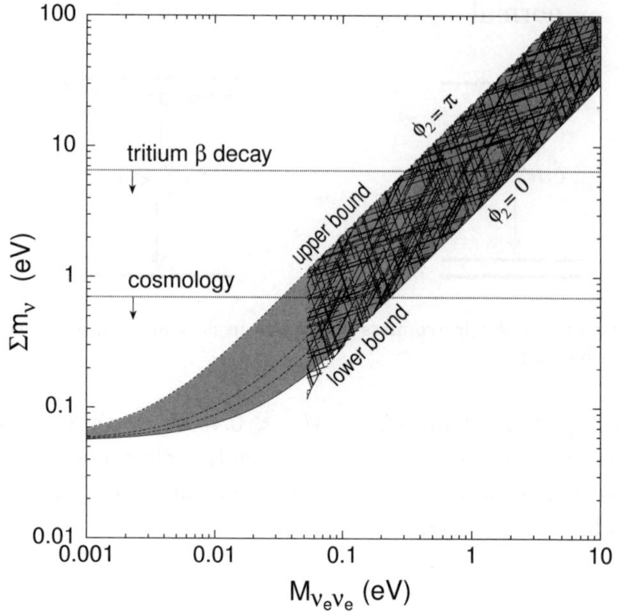

Fig. 7. Σ vs. M_{ee} for the normal (shaded) and inverted (cross-hatched) heirarchies. For the inverted hierarchy, $M_{ee} \geq \sqrt{\Delta}$, the 95% C.L. bounds from tritium β decay and cosmology are shown. Adapted from V. Barger, S.L. Glashow, D. Marfatia, and K. Whisnant, hep-ph/0201262, Phys. Lett. **B532**, 15 (2002).

nos, valuable for oscillation analyses, and lower backgrounds to ν_e appearance, for which we have presently only an upper bound.

The Grand Challenge is to test CP violation in the lepton sector. The critical parameters for this test are θ_x, sign(δm_a^2), which fixes the hierarchy of neutrino masses, and the CP-violating phase δ. Earth-matter effects are essential in this enterprise, both to determine sign(δm_a^2) and to resolve 8-fold parameter degeneracies that can confuse CP-violating and CP-conserving solutions $[(\theta_x, \delta); \text{sign}(\delta m_a^2); (\theta_a, \frac{\pi}{2} - \theta_a)]$. Long baselines are needed (>900 km) because the matter effects increase with distance and CP-violation effects require that the δm_s^2 oscillations contribute too.

There are two magic baselines for these studies. The first is the baseline for which the detector is located on the peak of the leading oscillation:

$$L = 495 \text{ km} \left(\frac{E}{1 \text{ GeV}}\right) \left(\frac{2.5 \times 10^{-3}\,\text{eV}^2}{\delta m_a^2}\right). \tag{15}$$

Then the $\nu_\mu \to \nu_e$ probability depends only on $\sin\delta$ (not $\cos\delta$) and this

Table 1. Present knowledge of neutrino parameters and future ways of improving this knowledge.

3-neutrino observables	Present knowledge	Future
$\sum m_\nu$	<0.7 eV	LSS, β-decay, $\beta\beta$
θ_a	$45° \pm 8°$	MINOS
θ_s	$33° \pm 3°$	SNO NC, KamLAND
θ_x	$\leq 9°$	$P(\nu_\mu \to \nu_e)$ LBL
δm^2_a	$\sim 2.5 \times 10^{-3}$ eV^2	MINOS
sign(δm^2_a)	unknown	$P(\nu_\mu \to \nu_e), P(\overline{\nu}_\mu \to \overline{\nu}_e)$ LBL
δm^2_s	$\sim 7.0 \times 10^{-5}$ eV^2	KamLAND
sign(δm^2_s)	+	
δ	unknown	$P(\nu_\mu \to \nu_e), P(\overline{\nu}_\mu \to \overline{\nu}_e)$ LBL
Majorana	unknown	$0\nu\beta\beta$
φ_2, φ_3	unknown	hopeless?

breaks the δ, θ_x degeneracy. The second magic baseline is $L = 7600$ km at which there is no δ dependence of the ν_e appearance probability.

Strategies for the future include Superbeams, with upgrades of the neutrino flux by a factor of 4 to 5, and long baselines. It is found that two Superbeam experiments do significantly better than one in parameter determinations. For example, the combination of data from a SJHF to SuperK experiment ($\theta_{\text{off-axis}} = 2$ deg, $L = 295$ km, 22.5 kt Water Cherenkov) and a SNuMI to Superior experiment (1 deg, $L = 900$ km, 2 kt Low-Z Calorimeter), both with 2 years ν_μ running and 6 years $\overline{\nu}_\mu$ running, would be sensitive to both the sign of δm^2_a and to CP violation at $\sin^2 2\theta_x > 0.03$. Another approach is to use a wide-band Superbeam, for example BNL to a Neutrino Underground Science Laboratory.

A Neutrino Factory (NuFact) is the ultimate technology for neutrino oscillation studies. Muons would be stored in an oval ring and their decays will give neutrino beams in the directions of the straight sections of the storage ring. Stored muons of energies above 20 GeV are needed and energies as high as 50 GeV have been considered in design studies. A Neutrino Factory would have sensitivity down to 10^{-4} and possibly lower in $\sin^2 2\theta_x$ for both sign(δm^2_a) and CP-violation determinations. A NuFact would also give the first electron-neutrino and electron-antineutrino beams. Precision reconstruction of the neutrino mixing matrix would be possible with a Neutrino Factory.

39. The Outlier: LSND

All of the above discussion was based on the assumption of 3 neutrinos. The LSND experiment found evidence for $\overline{\nu}_\mu \to \overline{\nu}_e$ oscillations at 4σ significance that cannot be accommodated by oscillations of three neutrinos, which allow only two distinct δm^2 scales. The LSND oscillation parameters are $\delta m^2_{\rm LSND} \sim 0.2$–$1\,{\rm eV}^2$ and $\sin^2 2\theta_{\rm LSND} \sim 0.02$. To explain the LSND effect a sterile neutrino has been invoked to have oscillations involving 4 neutrinos, in schemes with 2 + 2 and 3 + 1 mass hierarchies wherein the LSND scale is the large mass-scale separation. However, in global fits to all oscillation data these 4-neutrino schemes are at best borderline allowed. We wait for MiniBoone to confirm or reject with finality the LSND effect.

Acknowledgments

I thank Danny Marfatia and Kerry Whisnant for helpful comments on the contents of this talk. This research was supported in part by the National Science Foundation under Grant No. PHY99-07949, in part by the U.S. Department of Energy under Grant No. DE-FG02-95ER40896 and in part by the Wisconsin Alumni Research Foundation. I thank the Kavli Institute for Theoretical Physics at the University of California in Santa Barbara for hospitality.

References

For references to the literature on the topics addressed in this talk the reader is referred to the following recent reviews:

(1) S. Pakvasa and J.W.F. Valle, Neutrino Properties Before and After Kamland, hep-ph/0301061.
(2) M.C. Gonzalez-Garcia and Yosef Nir, Developments in Neutrino Physics, hep-ph/0202058.

Roman W. Jackiw

Proceedings of the Dirac Centennial Symposium
Howard Baer and Alexander Belyaev

Dirac's Magnetic Monopoles (Again)

Roman W. Jackiw
Center for Theoretical Physics
Massachusetts Institute of Technology
Cambridge, MA 02139-4307

Dirac's quantization of magnetic monopole strength is derived without reference to a (singular, patched) vector potential.

Dirac's monumental works opened various areas of inquiry in physics and mathematics. His equation not only became the paradigmatic description for the elementary constituents of matter, but also was recognized by mathematicians as encoding in its eigenvalues far-reaching information about geometry and topology of manifolds. His delta function stimulated the development of an entire field of mathematics – the theory of distributions, or generalized functions. This is how Laurent Schwartz, the creator of that field, put it:

> I heard of the Dirac function for the first time in my second year at the E[cole] N[ormale] S[upérieure].... those formulas were so crazy from the mathematical point of view that there was simply no question of accepting them.[1]

My own research, like that of all other physicists, is completely dependent on these magnificent explorations by Dirac. But there also are other paths that he blazed, which I have followed. He formulated the quantization of field theory on unconventional surfaces, corresponding to classical initial value problems posed on these surfaces. This suggested my construction (with Cornwall) of light-cone current algebra,[2] and (with Fubini and Hanson) of radial quantization,[3] which now is a tool in string theory. Dirac showed how to quantize dynamical systems that evolve in time while obeying constraints. Reformulating his approach, in order to simplify it, led Faddeev and me to propose a Darboux-based solution to the same problem.[4] Dirac posited a time-dependent variational principle, which leads to the time-dependent Schrödinger equation. Even though he didn't seem to publicize it – it appears only in an appendix to the Russian translation of his textbook – Kerman and I used it to define variationally the quantum effective action, and in an approximate implementation of the variational principle to derive the time-dependent Hartree-Fock equations.[5] These days his concept of a filled negative energy sea appears old-fashioned and awkward; mostly it is replaced by normal ordering prescriptions in quantum field theory. Nevertheless, reference to this apparently unphysical construct gives the most physical picture for quantum anomalies and for charge fractionation, as was demonstrated by Feynman[6] and Schrieffer,[7] respectively.

A particularly tantalizing result by Dirac concerns his monopoles. As is well known, he showed that within quantum mechanics monopole strength has to be quantized, but the quantization does not arise from a quantal eigenvalue problem. Rather quantization is enforced by the requirement that the phase-exponential of the classical action be gauge invariant. The Lagrangian and the action for motion in a magnetic field are not manifestly gauge invariant, since they involve the gauge-variant vector potential, rather than the gauge-invariant magnetic field. Moreover, because the vector potential for magnetic monopoles is singular, a gauge transformation shifts the action by a constant, and the phase exponential of the action remains unchanged only when this constant is a proper multiple of 2π. This then is the origin of Dirac's famous quantization condition, and it has a precise field theoretical reprise in the quantization of the Chern-Simons coefficient in odd-dimensional gauge theories, as was shown by Deser, Templeton, and me.[8]

Dirac's quantization argument has been thoroughly scrutinized, and is certainly acceptable. But one wonders whether one could reach the same conclusion in a gauge-invariant manner, relying on gauge-invariant quantities and dispensing with reference to gauge-variant and singular vector potentials.

Here I shall present such an argument, which I constructed some years ago.[9] Although it is not new, it is not widely known. Moreover, it not only regains the Dirac quantization condition, but also demonstrates that quantal magnetic sources must be structureless point particles.

Let us begin by recording the gauge-invariant Lorentz-Heisenberg equations of motion for operators $\boldsymbol{r}(t)$ specifying the motion of a massive (m) charged (e) particle in an external magnetic field $\boldsymbol{B}$:

$$\dot{\boldsymbol{r}} = \boldsymbol{v} \tag{1}$$

$$m\dot{\boldsymbol{v}} = \frac{e}{2c}\big[\boldsymbol{v} \times \boldsymbol{B} - \boldsymbol{B} \times \boldsymbol{v}\big] . \tag{2}$$

In the second equation, the noncommuting operators $\boldsymbol{v}$ and $\boldsymbol{B}(\boldsymbol{r})$ are symmetrized. Since the magnetic field does no work, the conserved energy $\frac{1}{2}m\boldsymbol{v}^2$ does not see it. This energy formula also gives us the Hamiltonian H that generates the above equations by commutation,

$$H = \frac{\boldsymbol{\pi}^2}{2m} \qquad \boldsymbol{\pi} \equiv m\boldsymbol{v} \tag{3}$$

provided the following brackets are posited:

$$[r^i, r^j] = 0 \tag{4}$$

$$[r^i, \pi^j] = i\hbar\delta^{ij} \tag{5}$$

$$[\pi^i, \pi^j] = ie\frac{\hbar}{c}\varepsilon^{ijk}B^k(\boldsymbol{r}) . \tag{6}$$

Note that π^i is not the (gauge-variant) canonical momentum; rather it is the (gauge-invariant) kinematical momentum. With (3)–(6) eqs. (1) and (2) are reproduced as

$$\dot{\boldsymbol{r}} = \frac{i}{\hbar}[H, \boldsymbol{r}] = \frac{\boldsymbol{\pi}}{m} \tag{7}$$

$$\dot{\boldsymbol{\pi}} = \frac{i}{\hbar}[H, \boldsymbol{\pi}] = \frac{e}{2mc}\big[\boldsymbol{\pi} \times \boldsymbol{B} - \boldsymbol{B} \times \boldsymbol{\pi}\big] . \tag{8}$$

The equations of motion (1), (2) or (7), (8) do not appear to require any constraint on $\boldsymbol{B}$. They make sense whether $\boldsymbol{B}$ is source free $\vec{\nabla} \cdot \boldsymbol{B} = 0$, or not $\vec{\nabla} \cdot \boldsymbol{B} \neq 0$. However, when we look to the Jacobi identity for the commutators of the $\boldsymbol{\pi}$'s, we find $\frac{1}{2}\varepsilon^{ijk}\big[\pi^i, [\pi^j, \pi^k]\big] = \frac{e\hbar^2}{c}\vec{\nabla} \cdot \boldsymbol{B}$. This vanishes, as it should, for source-free magnetic fields, which then are given by the conventional curl of a vector potential, $\boldsymbol{B} = \vec{\nabla} \times \boldsymbol{A}$, and momenta $\boldsymbol{p}$ canonically conjugate to $\boldsymbol{r}$ realize the algebra (5), (6) with the formula

$$\boldsymbol{\pi} = \boldsymbol{p} - \frac{e}{c}\boldsymbol{A}(\boldsymbol{r}) . \tag{9}$$

But how are we to understand the occurrence of magnetic sources with the concomitant violation of the Jacobi identities? To make progress on this question, recall that commutators in an algebra state the infinitesimal composition law for the corresponding finite transformations. In particular (5) shows that

$$T(\boldsymbol{a}) \equiv \exp\left(-\frac{i}{\hbar}\boldsymbol{a}\cdot\boldsymbol{\pi}\right) \tag{10}$$

effects translations by $\boldsymbol{a}$ on $\boldsymbol{r}$:

$$T^{-1}(\boldsymbol{a})\,\boldsymbol{r}\,T(\boldsymbol{a}) = \boldsymbol{r} + \boldsymbol{a}\,. \tag{11}$$

If the $\boldsymbol{\pi}$ were commuting momenta, the product of $T(\boldsymbol{a}_1)$ with $T(\boldsymbol{a}_2)$ would reflect the Abelian composition law of translations and close on $T(\boldsymbol{a}_1+\boldsymbol{a}_2)$. Here, however, because the $\boldsymbol{\pi}$'s do not commute [see (6)] we find

$$T(\boldsymbol{a}_1)T(\boldsymbol{a}_2) = \exp\Big(-\frac{ie}{\hbar c}\Phi(\boldsymbol{r};\boldsymbol{a}_1,\boldsymbol{a}_2)\Big)T(\boldsymbol{a}_1+\boldsymbol{a}_2) \tag{12}$$

where $\Phi(\boldsymbol{r};\boldsymbol{a}_1,\boldsymbol{a}_2)$ is the magnetic flux through the triangle with vertices $(\boldsymbol{r};\boldsymbol{r}+\boldsymbol{a}_1,\boldsymbol{r}+\boldsymbol{a}_1+\boldsymbol{a}_2)$ (in the direction $\boldsymbol{a}_1\times\boldsymbol{a}_2$). (See Fig. 1.)

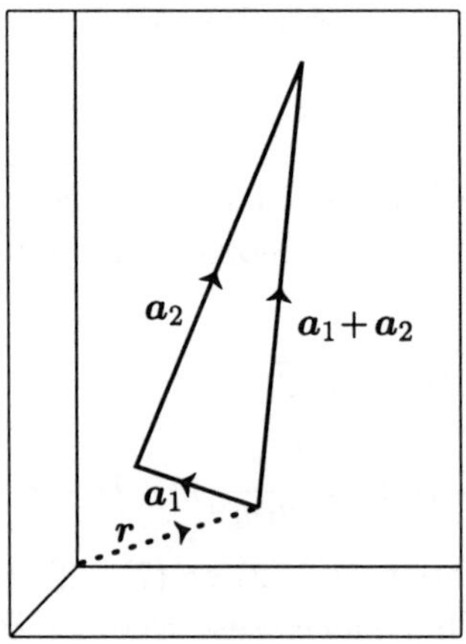

Fig. 1. The triangle at $\boldsymbol{r}$ through which the flux Φ is calculated.

The Jacobi identity is the infinitesimal statement of associativity in the composition law. Its failure when $\vec{\nabla}\cdot\boldsymbol{B}\neq 0$ means that finite translations do not associate. Indeed, from (12) we have, on the one hand,

$$\begin{aligned}(T(\boldsymbol{a}_1)T(\boldsymbol{a}_2))T(\boldsymbol{a}_3) &= \exp\left(-\frac{ie}{\hbar c}\Phi(\boldsymbol{r};\boldsymbol{a}_1,\boldsymbol{a}_2)\right)T(\boldsymbol{a}_1+\boldsymbol{a}_2)\,T(\boldsymbol{a}_3) \\ &= \exp\left(-\frac{ie}{\hbar c}\Phi(\boldsymbol{r};\boldsymbol{a}_1,\boldsymbol{a}_2)\right)\exp\left(-\frac{ie}{\hbar c}\Phi(\boldsymbol{r};\boldsymbol{a}_1+\boldsymbol{a}_2,\boldsymbol{a}_3)\right)T(\boldsymbol{a}_1+\boldsymbol{a}_2+\boldsymbol{a}_3)\end{aligned} \tag{13}$$

and on the other

$$\begin{aligned}T(\boldsymbol{a}_1)(T(\boldsymbol{a}_2)T(\boldsymbol{a}_3)) &= T(\boldsymbol{a}_1)\exp\Big(-\frac{ie}{\hbar c}\Phi(\boldsymbol{r};\boldsymbol{a}_2,\boldsymbol{a}_3)\Big)T(\boldsymbol{a}_2+\boldsymbol{a}_3) \quad (14)\\ &= \exp\Big(-\frac{ie}{\hbar c}\Phi(\boldsymbol{r}-\boldsymbol{a}_1;\boldsymbol{a}_2,\boldsymbol{a}_3)\Big)T(\boldsymbol{a}_1)T(\boldsymbol{a}_2+\boldsymbol{a}_3)\\ &= \exp\Big(-\frac{ie}{\hbar c}\Phi(\boldsymbol{r}-\boldsymbol{a}_1;\boldsymbol{a}_2,\boldsymbol{a}_3)\Big)\exp\Big(-\frac{ie}{\hbar c}\Phi(\boldsymbol{r};\boldsymbol{a}_2+\boldsymbol{a}_3)\Big)T(\boldsymbol{a}_1+\boldsymbol{a}_2+\boldsymbol{a}_3)\,.\end{aligned}$$

Putting everything together we find that

$$\Big(T(\boldsymbol{a}_1)T(\boldsymbol{a}_2)\Big)T(\boldsymbol{a}_3) = \exp\Big(-\frac{ie}{\hbar c}\omega(\boldsymbol{r};\boldsymbol{a}_1,\boldsymbol{a}_2,\boldsymbol{a}_3)\Big)T(\boldsymbol{a}_1)\Big(T(\boldsymbol{a}_2)T(\boldsymbol{a}_3)\Big) \quad (15)$$

where $\omega(\boldsymbol{r};\boldsymbol{a}_1,\boldsymbol{a}_2,\boldsymbol{a}_3)$ is the total magnetic flux emerging out of the tetrahedron formed from three vectors $\boldsymbol{a}_i$, with one vertex at $\boldsymbol{r}$:

$$\omega = \int \mathrm{d}\boldsymbol{S}\cdot\boldsymbol{B} = \int \mathrm{d}r\,\overrightarrow{\nabla}\cdot\boldsymbol{B}\,. \quad (16)$$

The last integral is over the interior of the tetrahedron, and of course vanishes for source-free magnetic fields, but is nonzero in the presence of magnetic sources, leading in general to the nonassociativity of the translations $T(\boldsymbol{a})$. (See Fig. 2.)

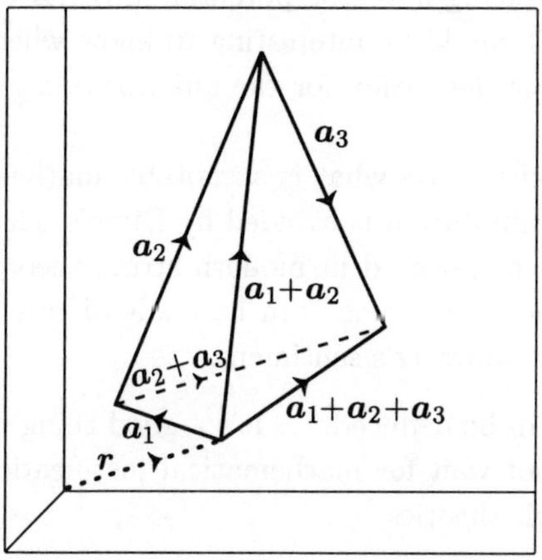

Fig. 2. The tetrahedron at $\boldsymbol{r}$ through which the flux determining the nonassociative phase ω is calculated.

But when operators act on a vector or Hilbert space, they necessarily associate. So one cannot tolerate nonassociativity within the usual quantum mechanical formalism. The only possibility for nonvanishing $\vec{\nabla} \cdot \boldsymbol{B}$ is that its integral is quantized for arbitrary $\boldsymbol{a}_1$, $\boldsymbol{a}_2$, and $\boldsymbol{a}_3$:

$$\int \mathrm{d}r\, \vec{\nabla} \cdot \boldsymbol{B} = 2\pi \frac{\hbar c}{e} N \,. \tag{17}$$

Then $\frac{e}{\hbar c}\omega$ is invisible in the exponent since it is an integer (N) multiple of 2π.

Equation (17) saves associativity, but it places requirements on $\boldsymbol{B}$. First, the magnetic field must be a (collection of) point source(s), so that (17) not lose its integrality when the $\boldsymbol{a}_i$ are varied: the source must be either inside or outside the tetrahedron. Moreover, for each point source of strength g

$$\boldsymbol{B} = g \frac{\boldsymbol{r}}{r^3} \tag{18}$$

we have $\vec{\nabla} \cdot \boldsymbol{B} = 4\pi g \delta^3(\boldsymbol{r})$ and Dirac's quantization is regained from (17):

$$\frac{ge}{\hbar c} = \frac{N}{2} \,. \tag{19}$$

Finally, with point sources we can also save the Jacobi identity, which now is violated only at isolated points and these may be excluded from the manifold.

Thus one has arrived at Dirac's result in a gauge-invariant manner, without ever introducing a vector potential with its attendant singularities, patches, etc. It would be interesting to know whether there is a similarly gauge-invariant derivation for the quantization of the Chern-Simons coefficient.[8]

Noncommutativity skirts what is acceptable mathematics for quantum theory. Its first manifestation is avoided by Dirac's quantization. Yet noncommutativity has reappeared in modern string theory. It remains to be seen whether mathematical sense can be made of this. Here again we can appreciate Laurent Schwartz's sentiment,

> This at least can be deduced.... It's a good thing that theoretical physicists do not wait for mathematical justification before going ahead with their theories.[1]

What about the physics, as opposed to the mathematics, of magnetic monopoles? Let me conclude with Dirac's own assessment:

> I am inclined now to believe that monopoles do not exist. So many years have gone by without any encouragement from the experimental side.[10]

References

1. L. Schwartz, *A Mathematician Grappling with His Century* (Birkhäuser, Basel 2001).
2. J. Cornwall and R. Jackiw, "Canonical Light-cone Commutators", *Phys. Rev. D* **4**, 367 (1971).
3. S. Fubini, A. Hanson, and R. Jackiw, "New Approach to Field Theory", *Phys. Rev. D* **7**, 1732 (1973).
4. L. Faddeev and R. Jackiw, "Hamiltonian Reduction of Unconstrained and Constrained Systems", *Phys. Rev. Lett.* **60**, 1692 (1988).
5. R. Jackiw and A. Kerman, "Time-dependent Variational Principle and the Effective Action", *Phys. Lett.* **71A**, 158 (1979).
6. R.P. Feynman (unpublished private communication).
7. R. Jackiw and J.R. Schrieffer, "Solitons with Fermion Number 1/2 in Condensed Matter and Relativistic Field Theories", *Nucl. Phys.* **B190 [FS3]**, 253 (1981).
8. S. Deser, R. Jackiw, and S. Templeton, "Topologically Massive Gauge Theories", *Ann. Phys. (NY)* **140**, 372 (1982).
9. R. Jackiw, "Three-Cocycle in Mathematics and Physics", *Phys. Rev. Lett.* **54**, 159 (1985); see also A.Z. Jadczyk, "Magnetic Charge Quantization and Generalized Imprimitivity Systems", *Int. J. Theor. Phys.* **14**, 183 (1975).
10. P.A.M. Dirac, letter to A. Salam, November 1981.

Joe Polchinski

Proceedings of the Dirac Centennial Symposium
Howard Baer and Alexander Belyaev

Monopoles, Duality, and String Theory

Joe Polchinski
Kavli Institute for Theoretical Physics,
University of California,
Santa Barbara, CA 93106-4030, USA
E-mail: joep@kitp.ucsb.edu

Dirac showed that the existence of magnetic monopoles would imply quantization of electric charge. I discuss the converse, and propose two 'principles of completeness' which I illustrate with various examples. Presented at the Dirac Centennial Symposium, Tallahassee, Dec. 6-7, 2002.

40. Theory

It is a great honor to speak at this centennial of Paul Dirac. I also had the honor to speak at Pauli's centennial two years ago and at Heisenberg's last year, and on each occasion it has been interesting to go back and learn more about the work of these men. Of course the high point of their scientific lives came rather early, with the discovery of quantum mechanics, and the rest was anticlimax by comparison. But from a modern perspective the latter part is also fascinating, as they went on to confront many problems that are still timely today.

All three thought very hard about the divergences of quantum field theory, which was perhaps the central theoretical question of the day. On the other hand, Pauli and Heisenberg both looked for unified theories, while Dirac regarded this as premature. It was his conviction that the immediate problem facing theoretical physicists was to develop better mathematical tools:[1]

> The most powerful method of advance [is] to perfect and generalize the mathematical formalism that forms the existing basis of theoretical physics.

Remarkably, not only did Dirac identify the necessary direction, he followed it successfully and provided many key ideas that continue to play a major role today:

- Magnetic monopoles.
- Path integrals.

- Light cone dynamics.
- Membrane actions.
- Conformal and de Sitter symmetries.
- Constrained Hamiltonian dynamics.
- Canonical formulation of gravity.

For an anticlimax that is a pretty good career.

Dirac comes across in many ways as the first modern theoretical physicist. Many of his statements illustrate this, but the following strikes me as particularly apt:[2]

> One must be prepared to follow up the consequences of theory, and feel that one just has to accept the consequences no matter where they lead.

Dirac is often quoted on the importance of mathematical beauty in one's equations; I did not choose one of these quotations because beauty is so difficult to define. He also made various statements that one should not be distracted by experiment; I did not choose one of these because they are inflammatory.

The reason that I find the chosen quotation so striking is that it is not supposed to be possible to follow theory alone. Without experimental guidance, it is said, one will quickly become lost. But of course today in high energy theory we are to a large extent following theory where it leads us, and we are rather confident that this is a correct and fruitful path. Why this approach can work is illustrated by Dirac's great discovery:

$$\text{quantum mechanics} + \text{special relativity} \;\Rightarrow\; \text{antiparticles}\,. \tag{1}$$

This was not a direct deduction (though in the framework of quantum field theory, one can show that antiparticles are necessary for causality.[3]). Rather, when Dirac tried to find a consistent framework that combined quantum theory and special relativity, he found it very difficult — so much so that when he did find one he had great confidence in its inevitability, and was prepared to take its other consequences seriously.

Essentially, with the discovery of quantum mechanics and special relativity, and even more so with general relativity, theory has become very rigid, so that it is difficult to extend or modify our existing theories without making them inconsistent or otherwise unattractive. One reason for this is that relativity unifies space and time. For example, it seems almost inevitable that in quantum gravity space will be modified or cut off at short distance. One can imagine many sorts of modification to the structure of space, but it is much harder to alter the nature of time in a consistent way.

41. The Necessity of Monopoles

One of Dirac's remarkable discoveries was the connection between magnetic monopoles and charge quantization.[1,4] Very early in the history of quantum theory, he recognized the important connection between geometry and quantum mechanics. Dirac showed that in the presence of a magnetic charge g, in order for the quantum mechanics of an electric charge e to be consistent one had to have

$$eg = 2\pi n \; . \tag{2}$$

Thus the existence of even a single magnetic charge forces every electric charge to be a multiple of $2\pi/g$. From the highly precise electric charge quantization that is seen in nature, it is then tempting to infer that magnetic monopoles exist, and indeed Dirac did so:[1]

> One would be surprised if nature had made no use of it.

I would like to discuss this from the point of view of the modern search for a unified theory, and to offer two general principles of completeness:

(1) In any theoretical framework that requires charge to be quantized, there will exist magnetic monopoles.

(2) In any fully unified theory, for every gauge field there will exist electric and magnetic sources with the minimum relative Dirac quantum $n = 1$ (more precisely, the lattice of electric and magnetic charges is maximal).

Obviously neither of these is a theorem. Rather, they are aesthetic principles based on experience with a rather wide range of examples. I will give three examples of the first principle, and two of the second.

41.1. *Grand Unification*

The most well-known example of the first principle of completeness is the 't Hooft-Polyakov monopole.[5] If the $U(1)$ of electromagnetism is embedded in a semisimple group, for example in grand unification

$$SU(3) \times SU(2) \times U(1) \subset SU(5) \; , \tag{3}$$

then electric charge is necessarily quantized, since it descends from the quantized representations of the unified group. Under precisely these conditions, 't Hooft and Polyakov showed that magnetic monopoles will exist as smooth but topologically nontrivial classical solutions.

Let me give a brief description of this idea. Dirac showed that the vector potential for a magnetic charge had a singularity along a string extending from the charge.

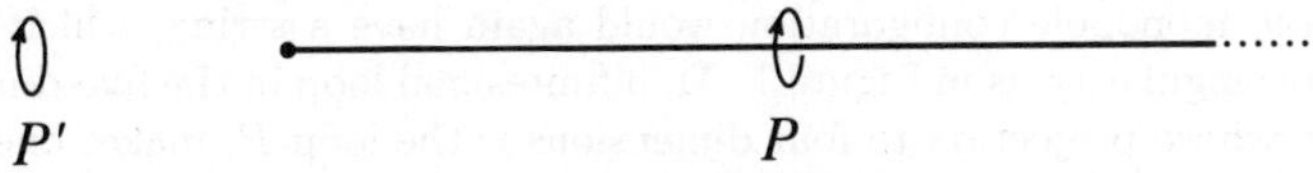

Fig. 1. A Dirac string extending from the monopole.

Figure 1 shows a Dirac string. Let us parallel transport a charged field ψ on the infinitesimal path P around it according to

$$d\psi = ie\boldsymbol{A} \cdot \boldsymbol{dx}\psi \; . \tag{4}$$

Because of the string singularity, the field picks up a net phase eg in the process; this is unobservable precisely if $eg = 2\pi n$. One can think of the phase of ψ as looping around the $U(1)$ group, which is a circle, n times. Now, if we pull the loop off of the string to the position P', the field is nonsingular and the phase is constant. Hence the net phase must drop rapidly from $2\pi n$ to zero as the loop is pulled past the monopole, and this observable phase signifies a singularity in the field. When $U(1)$ is embedded in a semisimple group, however, the loop can become topologically trivial: in $SU(2)$ for example, a rotation through 4π can be smoothly deformed to a trivial path. Thus there are smooth field configurations, which at long distance look like Dirac monopoles, with a net $U(1)$ magnetic charge.

As a postscript to the talk, I should note that this argument does not require that one can obtain $eg = 2\pi$, and indeed this depends on the field content. In the Georgi-Glashow model $U(1) \subset SU(2)$, if there are only fields of integer isospin then the minimum quantum is $eg = 4\pi$, while if there are fields of half-integer isospin then $eg = 2\pi$ is obtained. (This is related to the fact that some simple groups do have a finite set of nontrivial closed loops.) So in the sense of the second principle I would have to say that this theory is not fully unified, precisely because its matter content is not fixed but subject to arbitrary choice.

41.2. *Kaluza-Klein Theory*

If spacetime is five-dimensional, with the added dimension x^4 being periodic, then five-dimensional gravity gives rise to both gravity and a Maxwell field in four dimensions. This was perhaps the first application of spontaneous symmetry breaking as a unifying concept — the laws of physics are invariant under Lorentz transformations in all five spacetime dimensions, while the state we live in is invariant only under the four-dimensional symmetry group. The metric components $g_{\mu 4}$ become the Maxwell potential, and gauge invariance arises from reparameterizations of x^4. What the four-dimensional physicist sees as electric charge is therefore momentum in the 4-direction, and it is quantized because of the periodicity in this direction.

A Dirac monopole configuration would again have a string, which is now a coordinate singularity as in Figure 1. An infinitesimal loop in the five-dimensional geometry, whose projection to four dimensions is the loop P, makes one or more circuits of the x^4 coordinate. An infinitesimal loop away from the string, such as P', does not loop the x^4 direction. Thus there is again singular behavior as the loop is pulled past the monopole. However, Gross, Perry, and Sorkin[6] showed that again there are smooth geometries that look like Dirac monopoles outside of some core region. The point is that if the radius of the x^4 direction shrinks to zero in an appropriate way, then at the origin there is only a coordinate singularity and the loop can be smoothly slid off the string.

41.3. *U*(1) *Lattice Gauge Theory*

The final example of the first principle is rather different from the others, but very vividly illustrates the connection between charge quantization and magnetic monopoles. When one puts $U(1)$ gauge theory on a spatial or spacetime lattice, the basic variables A_l live on links, directed pairs of adjacent points. One can think of A_l as the lattice approximation to $\int \boldsymbol{A} \cdot d\boldsymbol{x}$, integrated from one site to the next.

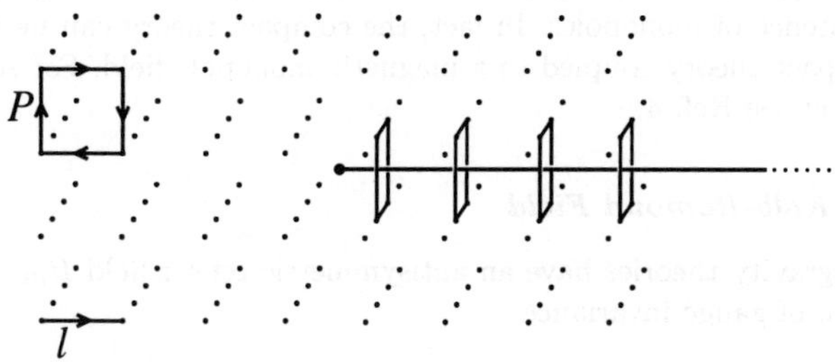

Fig. 2. Part of a spatial lattice. A typical link l and plaquette P are illustrated. A magnetic monopole and its Dirac string are shown, hidden between the sites of the lattice.

There are two versions of $U(1)$ lattice gauge theory, according to whether A_l is a periodic variable or takes values on the whole real line:

$$\begin{aligned} &\text{compact theory:} \quad A_l \cong A_l + 2\pi \ , \\ &\text{noncompact theory:} \quad -\infty < A_l < \infty \ . \end{aligned} \tag{5}$$

For example these theories have different actions. The analog of the field strength is the sum of the link variables around a plaquette,

$$F_P = \sum_{l \in P} A_l \ . \tag{6}$$

In the noncompact case the simplest action would be the sum over plaquettes of F_P^2, while in the compact case it would be the sum of $(1 - \cos F_P)$. Alternately, one can describe the compact theory entirely in terms of $U_l = e^{iA_l}$, and the action is constructed from $U_P = \prod_{l \in P} U_l$.

In the noncompact case charge is not quantized, one can transport any charge e from one site to the next with the phase e^{ieA_l}. However, in the compact case this is defined only for e an integer, and so charge is quantized (we are working here in units where the charge is dimensionless and the minimum value is 1, but we can rescale A_l to other systems of units). Precisely in keeping with the general principle, one finds that there are magnetic monopoles in the compact case but not the noncompact one. The Dirac string is a line of plaquettes on all of which $F_P \approx -1$ (or $-n$), while it is $\ll 1$ on all other plaquettes except those very close to the monopole. The line ends at the monopole. From the definition of F_P it follows that F_P summed over any closed surface is zero, so if we consider a large surface surrounding the monopole the sum of all the small fluxes must be $+1$ (or $+n$) to cancel the contribution of the string: this is the monopole flux. Nothing is singular, everything is made finite by the lattice. In the compact theory the string costs no energy because U_P is everywhere near unity, while in the noncompact theory the string is visible to noninteger charges and has an energy proportional to its length.

This example makes vividly clear the connection between charge quantization and the existence of monopoles. In fact, the compact theory can be rewritten as the noncompact theory coupled to a magnetic monopole field. For some reviews of this subject see Ref. 4.

41.4. *The Kalb-Ramond Field*

Many supergravity theories have an antisymmetric tensor field $B_{\mu\nu}$, with a generalized form of gauge invariance

$$\delta B_{\mu\nu} = \partial_\mu \lambda_\nu - \partial_\nu \lambda_\mu \ . \tag{7}$$

If we consider such a theory in a Kaluza-Klein geometry, the components $B_{\mu 4}$ again become a four-dimensional Maxwell theory. However, unlike the Maxwell field from the metric, there are no states in supergravity that are electrically charged under this gauge field — there is no way to minimally couple any field to $B_{\mu\nu}$ so as to give rise to a minimal coupling to $B_{\mu 4}$.

This incompleteness is not an inconsistency, but it is somewhat puzzling and unattractive that there is this asymmetry between the gauge fields from the metric and those from $B_{\mu\nu}$. In this respect string theory completes supergravity. The two-form $B_{\mu\nu}dx^\mu dx^\nu$ can be integrated over the world-sheet of the string, just as $A_\mu dx^\mu$ can be integrated over the world-line of a particle, and just such a coupling is present in string theory. If a string wraps around the periodic x^4 direction, the integral dx^4 produces a coupling to $B_{\mu 4}$. Thus the string winding states are electrically charged.

There should also be a corresponding magnetic source. We can discuss this in four dimensions, but it is clearer to start in the full ten-dimensional theory. In general a p-dimensional object (brane) couples to a $(p+1)$-form potential, through the integral

$$\int B_{\mu_1 \dots \mu_{p+1}} dx^{\mu_1} \dots dx^{\mu_{p+1}} \tag{8}$$

over the world-history of the brane. The curl of this potential gives a $(p+2)$-form field strength. Contracting this with the spacetime ε tensor gives a dual $(D-p-2)$-form field strength, where D is the dimension of spacetime. This corresponds to a $(D-p-3)$-form magnetic potential, and so couples magnetically to a p'-dimensional object for $p' = D-p-4$. In other words, $p+p' = D-4$. The familiar electric and magnetic charges are simply $p = p' = 0$ in $D = 4$, but the Dirac quantization argument extends directly to all such pairs.[7,8]

For the case at hand $D = 10$ and $p = 1$ and so $p' = 5$: the magnetic object is a five-brane. This was found by Callan, Harvey, and Strominger[9] as a solitonic solution to the low energy field theory of string theory — it is the Neveu-Schwarz (NS) five-brane. Curiously their first paper found a five-brane charge of 8 Dirac units. This is consistent, but rather odd. Shortly afterward the authors realized that they had used inconsistent normalizations of $B_{\mu\nu}$, and found that the magnetic charge is the minimum Dirac quantum.

41.5. *D-Branes*

The final example has great significance for me. In string theory, in addition to the NS-NS field $B_{\mu\nu}$ there are other 'Ramond-Ramond' form potentials. The terminology refers to the fact that a closed string state is the product of the states of its right- and left-moving oscillations, so one can make a bosonic field out of two bosonic states (these are called NS-NS) or out of two fermionic states (called R-R). Unlike $B_{\mu\nu}$ these forms do not couple to the fundamental string.

However, string theory has extended objects of a distinctive type, the D-branes. These are like topological defects, with the notable property that a (normally closed) string can end on them. There were various early discussions of strings with such fixed endpoints. In particular it was argued that these objects had to be included in string theory because one encountered them if one took the normal open string theory and followed it to small radius (T-duality). For some time afterward these objects were still regarded as curiosities, but in the wake of string duality, in which R-R charges are necessary to fill out the multiplets, it was realized[10] that the consistency of weak/strong duality and T-duality required that the D-branes carry the R-R charge. Further, they exist with just the right dimensions to provide a full set of electric and magnetic sources, even for nondynamical 9- and 10-form fields.

The calculation of eg is interesting. It is shown schematically in Figure 3.

This process can be regarded as the emission and absorption of a closed string, giving rise to a force between the two D-branes. It can equally be regarded as a vacuum loop of an open string, and this is by far the most direct way to calculate it, by summing the zero point energies of open string modes. From the sum one reads off the potential, which gives the coupling of the D-branes to the various forms. Note that unlike the calculations of the charges of the various other objects that have been discussed there is nothing obviously topological about this, it is essentially the calculation of a Casimir energy. In particular there is no reason that it should give anything like an integer for eg. Indeed, the first attempts at the calculation gave undesirable powers of π and $1/2$, but after a day of debugging the result is that the product of the charge of a Dp-brane and that of a D$(6-p)$-brane is exactly 2π.

As a postscript, I should note that in any theory with gravity one can make electric and magnetic sources trivially, by having the field lines end on a black hole singularity (the R-R charged objects were first described in this form[11]). In this description there is no direct way to determine the actual spectrum of charges. However in string theory we can usually turn off the gravitational force

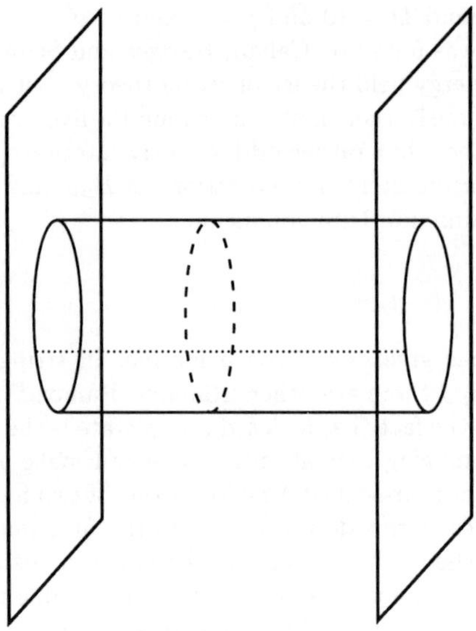

Fig. 3. Two D-branes (the vertical planes) and a cylindrical string world-sheet with one boundary on each D-brane. The world-sheet can be regarded either as an open string with one end on each D-brane (the dotted line) traveling in a vacuum loop, or a closed string (the dashed line) emitted by one D-brane and absorbed by the other.

by turning down the coupling, so that the black hole goes over to something nonsingular, and in this way we know the spectrum.

42. Conclusions

42.1. *The Existence of Monopoles*

By the end of his career Dirac became less certain about the existence of monopoles. He forgot his earlier dictum to ignore experiment! But as I have discussed, the existence of magnetic monopoles seems like one of the safest bets that one can make about physics not yet seen. It is very hard to predict when and if monopoles will be discovered. If their mass is at the grand unified scale as one expects, then they will be beyond the reach of accelerators, while inflation has almost certainly diluted any primordial monopoles beyond discovery. It is curious to contemplate this unfortunate situation, where theory predicts the existence of an object (and its production, but in experiments that can only be carried out in thought) and at the same time suggests that it may never be seen. But we must continue to hope that we will be lucky, or unexpectedly clever, some day.

42.2. *Duality and Beauty*

Many of the electric and magnetic objects that I have discussed look quite different from one another, but there is strong evidence that in each case that there are dualities that interchange them. For Dirac this would have been a triviality. He notes[4] that his theory of pointlike electric and magnetic charges is invariant under the interchange of the two objects, along with the interchange of electric and magnetic potentials. Dirac's theory is rather formal, since the magnetic coupling is the inverse fine structure constant, but one can regard the lattice as providing a precise definition of a cutoff theory, and with appropriate choice of the action it is self-dual.[7]

For the grand unified and Kaluza-Klein cases, however, any duality must be quite nontrivial. In these cases the electric charges are pointlike quanta, while the magnetic charges are smooth classical configurations. To be precise, this is the picture at weak coupling. Now it would not be surprising that as the coupling is turned up the electric objects begin to emit pairs and become big and fuzzy like the solitons. The great surprise (duality) is that when the coupling becomes very large the magnetic objects become more and more pointlike and the theory can be described in terms of their local fields. It is a remarkable property of the quantum theory that the degrees of freedom can, at least with some assistance from supersymmetry, reorganize themselves in this way. Indeed, we do not fully understand the details of this, but the number of independent consistency checks is enormous.

Even further, in string theory *all* of the electric and magnetic objects that have been discussed here, with the (possible[12]) exception of the lattice examples, are related to one another by dualities. These examples involve widely different aspects of gauge field geometry, spacetime geometry, 'stringy' geometry, string perturbation theory, and quantum and classical physics. The existence of a single structure that unifies such a broad range of physical and mathematical ideas, and many others as well, is unexpected and remarkable. Earlier I declined to define beauty, but one can recognize it when one sees it, and here it is. This is one illustration of why the scientific path that Dirac laid out has been such a fruitful one in recent times.

Acknowledgments

This work was supported by National Science Foundation grants PHY99-07949 and PHY00-98395.

References

1. P. A. Dirac, "Quantised Singularities In The Electromagnetic Field," Proc. Roy. Soc. Lond. A **133**, 60 (1931).
2. P. A. Dirac, 1977 Varenna lecture, quoted by M. Jacob in A. Pais, M. Jacob, D. I. Olive, and M. F. Atiyah, "Paul Dirac: The Man and his Work," Cambridge, UK: Univ. Pr. (1998).

3. See for example Chapter 2 of M. E. Peskin and D. V. Schroeder, "An Introduction To Quantum Field Theory," Reading, USA: Addison-Wesley (1995).
4. P. A. Dirac, "The Theory Of Magnetic Poles," Phys. Rev. **74**, 817 (1948).
5. G. 't Hooft, "Magnetic Monopoles In Unified Gauge Theories," Nucl. Phys. B **79**, 276 (1974);
   A. M. Polyakov, "Particle Spectrum In Quantum Field Theory," JETP Lett. **20**, 194 (1974) [Pisma Zh. Eksp. Teor. Fiz. **20**, 430 (1974)].
6. D. J. Gross and M. J. Perry, "Magnetic Monopoles In Kaluza-Klein Theories," Nucl. Phys. B **226**, 29 (1983);
   R. d. Sorkin, "Kaluza-Klein Monopole," Phys. Rev. Lett. **51**, 87 (1983).
7. M. E. Peskin, Annals Phys. **113**, 122 (1978);
   R. Savit, "Duality In Field Theory And Statistical Systems," Rev. Mod. Phys. **52**, 453 (1980).
8. P. Orland, "Instantons And Disorder In Antisymmetric Tensor Gauge Fields," Nucl. Phys. B **205**, 107 (1982);
   R. I. Nepomechie, "Magnetic Monopoles From Antisymmetric Tensor Gauge Fields," Phys. Rev. D **31**, 1921 (1985);
   C. Teitelboim, "Monopoles Of Higher Rank," Phys. Lett. B **167**, 69 (1986).
9. C. G. Callan, J. A. Harvey and A. Strominger, "Worldbrane Actions for String Solitons," Nucl. Phys. B **367**, 60 (1991); "World Sheet Approach To Heterotic Instantons And Solitons," Nucl. Phys. B **359**, 611 (1991).
10. J. Polchinski, "Dirichlet-Branes and Ramond-Ramond Charges," Phys. Rev. Lett. **75**, 4724 (1995) [arXiv:hep-th/9510017].
11. G. T. Horowitz and A. Strominger, "Black Strings and P-Branes," Nucl. Phys. B **360**, 197 (1991).
12. S. Hellerman, "Lattice Gauge Theories have Gravitational Duals," arXiv:hep-th/0207226.

Paul Langacker

Proceedings of the Dirac Centennial Symposium
Howard Baer and Alexander Belyaev

Time Variation of Fundamental Constants as a Probe of New Physics

Paul Langacker

Department of Physics and Astronomy
University of Pennsylvania
Philadelphia, PA 10104
E-mail: pgl@electroweak.hep.upen.edu

Time variation of fundamental constants would not be surprising in the framework of theories involving extra dimensions. The variation of any one constant is likely to be correlated with variations of others in a pattern that is diagnostic of the underlying physics.

43. Introduction

There has recently been reported evidence for a possible time variation of the fine structure constant on cosmological time scales.[1] Such variations are not surprising in any theoretical framework for the unification of basic forces involving extra dimensions or in which dimensionless couplings are related to the expectation values of scalar fields. However, the variation of α is likely to be correlated with the variations in other fundamental quantities, such as other gauge and Yukawa couplings, and the ratios of such dimensionful scales as the unification and electroweak or supersymmetry-breaking scales, or the unification and gravity scales. Thus, the observation of such variations is a powerful probe of the underlying physics. I briefly summarize the relevant issues and describe an analysis and parametrization of these effects done in collaboration with Matt Strassler and Gino Segrè.[2]

44. Theoretical motivations

There have been speculations going back to the pioneering work of Dirac in 1937 that the fundamental "constants" of nature may vary in time.[3,4,5,6,7] From a modern perspective, time-variation is not surprising. For example, in superstring theories and many brane-world scenarios, couplings are associated with moduli (scalar fields), which could be time-varying. In fact, time variation could be expected in any theory in which some or all of the couplings are associated with the expectation values of scalar fields,[8,9] provided that they vary on cosmological time scales. In the standard model, for example, masses are proportional to the expectation value of the Higgs field. Gauge and Yukawa couplings can similarly be associated with the expectation values of scalar fields that occur in higher-dimensional operators.

As a simple example, suppose there is a higher-dimensional operator coupling a scalar φ to the electromagnetic tensor $F_{\mu\nu}$,

$$\mathcal{L}_{\text{elm}} \sim \frac{1}{4}\left[1 + \frac{\lambda\varphi}{M_{\text{PL}}}\right] F_{\mu\nu}F^{\mu\nu} + \cdots, \tag{1}$$

where λ is dimensionless and M_{PL} is the Planck scale. It is useful to then replace A_μ by A'_μ, where

$$A_\mu = A'_\mu \left(1 - \frac{\lambda\varphi}{2M_{\text{PL}}}\right), \tag{2}$$

so that A'_μ has a canonical kinetic energy. The couplings of charged particles to A'_μ will then be canonical in terms of a rescaled electric charge e', related by

$$e = e'\left(1 + \frac{\lambda\varphi}{2M_{\text{PL}}}\right). \tag{3}$$

e' is universal, i.e., the rescaling is the same for all charged particles. If φ were a constant classical field, then the effects of these rescalings would be unobservable. However, if φ varies with time or in space, the effective electric charge e' would also vary. For example, if φ is time dependent, it would satisfy

$$\ddot{\varphi} + 3H\dot{\varphi} + \frac{\partial V}{\partial\varphi} = 0, \tag{4}$$

where H is the Hubble expansion rate and V is the scalar potential. φ could be associated with a field introduced for other purposes, e.g., quintessence,[10] or it might have no other cosmological significance (i.e., φ might or might not contribute significantly to H).

In addition to the time/space variation, there would be new operators associated with the derivatives of φ,[11] which are usually assumed to be small for small variations[a]. There would also be new long-range forces coupling to electromagnetic energy density mediated by the quantum of φ. These would violate the equivalence principle and could lead to strong but model-dependent bounds.[14,10]

One objection to the notion of time varying couplings is that in many frameworks the natural scale for the rate of variation of, e.g., the fine structure constant α, might be expected to be

$$\dot{\alpha}/\alpha \sim M_{\text{PL}} \sim 10^{+43}s^{-1}, \tag{5}$$

[a]Bekenstein has recently argued that they might in fact be relevant to the Webb et al. observations [12]. Implications for Lorentz and CPT violation are emphasized in [13].

while any actual variation is clearly very much smaller than this[b]. For example, the Webb et al. results suggest

$$\dot{\alpha}/\alpha \sim 10^{-15} yr^{-1} \sim 10^{-66} M_{\rm PL}. \tag{6}$$

It is tempting to assume that since $\dot{\alpha}/\alpha$ is so small compared to its natural scale it must be exactly zero or at least unobservably small for some reason. However, it is worth considering an analogy with the cosmological constant: in most frameworks the natural scale for the vacuum energy density, related to the cosmological constant by $\rho_{\rm vac} = \Lambda_{\rm cosm}/8\pi G_N$, is $\rho_{\rm vac} \sim M_{\rm PL}^4$. Most people assumed that since $\rho_{\rm vac}$ is so much smaller than this, there must be some principle to ensure $\rho_{\rm vac} = 0$. Recently, however, the Type IA supernova and CMB data have independently indicated that

$$\rho_{\rm vac} \sim 10^{-124} M_{\rm PL}^4 \neq 0. \tag{7}$$

(The observed dark energy may not be a true cosmological constant. It could be a time-varying quantity such as quintessence. For the purposes of this remark it does not make any difference.)

If α does vary with time, then it is likely that other fundamental constants, such as other gauge couplings α_i, Yukawa couplings h, the electroweak scale v, and the Newton constant $G_N = 1/M_{\rm PL}^2$ also vary[c] in a correlated way [2,4,25,26]. The relation of these quantities is presumably specified in any complete unified description of nature, though the form of the relations depends on the theory. One should therefore allow for the possibility that other quantities are varying when interpreting the observational data. Observations (or non-observations) of time or space variations can therefore be viewed as a probe of the underlying physics and how the various quantities are related.

45. Search for varying α

Webb et al. [1] have studied the absorption of light from background quasars by molecular clouds in the redshift range $0.5 < z < 3.5$. They apply a new "many multiplet" method to simultaneously study many relativistic (i.e., $O(\alpha^2, \alpha^4)$) splittings, obtaining evidence for an increase in α,

$$\frac{\Delta\alpha}{\alpha} \equiv \frac{\alpha_z - \alpha}{\alpha} = -(0.72 \pm 0.18) \times 10^{-5}, \tag{8}$$

where α_z (α) refers to the fine structure constant at redshift z (at present). This would correspond to $\dot{\alpha}/\alpha \sim 10^{-15}$/yr for $\dot{\alpha}/\alpha$ = constant. Using a different

[b]This is reminiscent of the flatness problem, expressed as the statement that the natural time scale for the evolution of the universe is $1/M_{\rm PL}$ rather than 1.4×10^{10} yr.

[c]I will take the view that only dimensionless couplings and ratios of mass scales are physically meaningful, and that quantities such as $\hbar$ and c are derived quantities rather than fundamental. In that case, they can be taken to be fixed at unity. For a debate on such matters, see [17].

method, Bahcall et al. [18] find a result consistent with no variation, though with lower precision, $\frac{\Delta\alpha}{\alpha} = (-2\pm1.2)\times10^{-4}$, for the redshift range 0.16-0.80. Similarly, Cowie and Songaila [19] constrain $X \equiv \alpha^2 g_p m_e/M_p$, where the proton magnetic moment is $eg_p/2M_p$, from the 21 cm hyperfine line in hydrogen at $z \sim 1.8$, and Potekhin *et al.* [20] limit $Y \equiv M_p/m_e$ from molecular hydrogen clouds at $z = 2.81$:

$$\frac{\Delta X}{X} = (0.7 \pm 1.1) \times 10^{-5}, \quad \frac{\Delta Y}{Y} = (8.3^{+6.6}_{-5.0}) \times 10^{-5}. \tag{9}$$

There are also stringent laboratory limits.[d] For example, Prestage et al. obtain [21] $|\frac{\Delta\alpha}{\alpha}| < 1.4\times10^{-14}$ over 140 days, corresponding to $\dot{\alpha}/\alpha < 3.7\times10^{-14}$/yr if constant. More recently, Sortais et al. obtained [22] $\dot{\alpha}/\alpha < (4.2\pm6.9)\times10^{-15}$/yr. Laboratory techniques may ultimately be sensitive to $\dot{\alpha}/\alpha < 10^{-18}/T$, where T is the running time [5].

A very stringent limit comes from the OKLO natural reactor [5]. In particular, the $^{149}Sm/^{147}Sm$ ratio is depleted by the capture of thermal neutrons,

$$n +^{149} Sm \rightarrow^{150} Sm + \gamma. \tag{10}$$

The cross section is dominated by a very low energy resonance, involving an almost exact cancellation between Coulomb and strong effects. Thus, even a small change in α could be significant. This was analyzed by Damour and Dyson [23] who found that $\dot{\alpha}/\alpha$ is bounded to be between -6.7×10^{-17}/yr and $+5.0\times10^{-17}$/yr, and by Fujii et al.,[24] who obtained $\dot{\alpha}/\alpha = (-0.2 \pm 0.8) \times 10^{-17}$/yr, both over 2×10^9 yr. This is a very stringent result, but does not directly contradict (8) because the latter refers to an earlier time period (around $(6 - 11) \times 10^9$ yr ago). Furthermore, only the possible variation in α was considered in.[23,24] It is conceivable that the effects of varying α could have been cancelled by a change in the strong interaction strength, α_s.

Big Bang Nucleosynthesis, which occurred for redshift $\sim 10^9 - 10^{10}$, implies [25] that $\frac{\Delta\alpha}{\alpha} < O(10^{-2})$, assuming that only α varies. This is weak compared with (8) if $\dot{\alpha}/\alpha$ is constant in time, but could conceivably be important if there were significantly enhanced effects at large redshift.

CMB results may eventually be able to constrain $\frac{\Delta\alpha}{\alpha}$ at the $10^{-2} - 10^{-3}$ level for $z \sim 1000$ from their effects on the ionization history of the Universe [26].

46. Correlations with α_s, h, v, $G_N, \cdots$

If α varies with time, it is likely that other fundamental constants do also. The correlations of their time dependences would be a probe of the underlying theory of particle physics [2,4,25,26].

For example, the observed low energy gauge couplings are consistent with the unification of the running gauge couplings at a scale $M_G \sim 3 \times 10^{16}$ GeV,

[d] Laboratory limits are reviewed in detail in [5].

predicted in simple supersymmetric grand unification [27]:

$$\frac{1}{\alpha_i(M_Z)} = \frac{1}{\alpha_G} + b_i t_G \tag{11}$$

where $\alpha_i, i = 1, 2, 3$ are the gauge couplings associated with $U(1) \times SU(2) \times SU(3)$, $t_G = \frac{1}{2\pi} \ln \frac{M_G}{M_Z} \approx 5.32$, $\alpha_G^{-1} \approx 23.3$ is the inverse of the common coupling at the unification scale, and the b_i are the beta function coefficients. In the MSSM, $b_i = \left(\frac{33}{5}, 1, -3\right)$. The (running) electromagnetic fine structure constant is related by $\alpha^{-1} = \frac{5}{3}\alpha_1^{-1} + \alpha_2^{-1} \sim 127.9$, where all three couplings are evaluated at M_Z. If gauge unification holds, either in the simple MSSM framework or something similar, then it is likely that all three gauge couplings will vary simultaneously.[2,15]

The simplest possibility is that the dominant effect is a time variation in α_G^{-1}. In that case, it is straightforward to show[2] that the strong coupling $\alpha_s = \alpha_3$ has a magnified variation,

$$\frac{\Delta\alpha_S}{\alpha_S} \simeq \frac{3}{8}\frac{\alpha_S}{\alpha}\frac{\Delta\alpha}{\alpha} \sim 5.8\frac{\Delta\alpha}{\alpha} \tag{12}$$

where α_s is evaluated at M_Z and we ignore the difference in the relative variation of α between scales 0 and M_Z. There is an even stronger variation in the QCD scale $\Lambda_{\rm QCD}$, at which α_s becomes strong,

$$\frac{\Delta\Lambda_{QCD}}{\Lambda_{QCD}} \sim 34\frac{\Delta\alpha}{\alpha}, \tag{13}$$

which is around -25×10^{-5} for the Webb et al. value (8). This has a theoretical uncertainty (given the assumptions) of around 20%. Most hadronic mass scales (with the exception of the pion mass) are approximately proportional to $\Lambda_{\rm QCD}$, so they are expected to have the same relative variation.

It is also reasonable to consider a variation in the electroweak scale $v \sim 246$ GeV (which sets the scale for $M_Z = g_W v$, where $\alpha_W = g_W^2/4\pi = 3\alpha_1/5 + \alpha_2$), or more precisely in the ratio of v to the unification scale M_G.[2,4,15,16] (Only dimensionless ratios of mass scales are physically relevant, so we are implicitly measuring all masses with respect to M_G.) In[2] we define the phenomenological parameter κ by

$$\frac{\Delta v}{v} \equiv \kappa\frac{\Delta\alpha}{\alpha}, \tag{14}$$

which implies that

$$\begin{aligned} \frac{\Delta\alpha_S}{\alpha_S} &\sim \frac{3}{8}\frac{\alpha_S}{\alpha}\left(1 - \frac{10\alpha}{\pi}\kappa\right)\frac{\Delta\alpha}{\alpha} \\ \frac{\Delta\Lambda_{QCD}}{\Lambda_{QCD}} &\sim 34\,(1 + 0.005\kappa)\,\frac{\Delta\alpha}{\alpha}. \end{aligned} \tag{15}$$

These corrections are small for κ of order unity, but important for larger κ. In fact, it is shown in[2] that $\kappa \sim 70$ in theories in which v is tied to the scale of soft supersymmetry breaking, and in which supersymmetry breaking occurs in a hidden sector at a scale in which a (unified) gauge coupling becomes strong! Even in this case, the correction to the Λ_{QCD} variation is only a factor of 1.35.

It is useful to introduce phenomenological parameters for the variation of other fundamental "constants". In particular, the variation of the Yukawa coupling h_a for fermion a (so that its Higgs-generated mass is $m_a = h_a v$) is parametrized as[e]

$$\frac{\Delta h_a}{h_a} \equiv \lambda_a \frac{\Delta \alpha}{\alpha}. \tag{16}$$

Similarly, the variation of the Planck scale $M_{PL} = G_N^{-1/2}$ (again, only the ratio of M_{PL} to other masss scales is relevant) is parametrized as

$$\frac{\Delta M_{PL}}{M_{PL}} \equiv \rho \frac{\Delta \alpha}{\alpha}. \tag{17}$$

The possible variation of various observables can be expressed in terms of these parameters, and their values can in principle be computed in any complete fundamental theory, allowing for a more general treatment of time variation[f]. For example, for the quantities defined before (9) one predicts[g] the variations,[2]

$$\begin{aligned} \frac{\Delta X}{X} &\sim (-32 + \lambda + 0.8\kappa) \frac{\Delta \alpha}{\alpha} \sim (23 \pm 6) \times 10^{-5} \\ \frac{\Delta Y}{Y} &\sim (34 - \lambda - 0.8\kappa) \frac{\Delta \alpha}{\alpha} \sim (-24 \pm 6) \times 10^{-5}, \end{aligned} \tag{18}$$

where I have assumed a common value λ for all the Yukawa factors λ_a, and the numerical values are evaluated using $\lambda = \kappa = 0$ and the Webb et al. value (8). These are to be compared with the experimental results in (9). Clearly, within this framework the observational results in (8) and (9) are consistent only if there is a delicate cancellation of effects, with $\lambda + 0.8\kappa \sim 32$. Other applications, including big bang nucleosynthesis, the OKLO reactor constraints, and the triple α process, are considered in.[2,4,15,16]

47. Conclusions

- Time (or space) variation of fundamental "constants" is plausible in any theory in which they are dependent on the sizes or properties of extra dimensions, or on other scalar fields.
- The natural scale for such variations in many frameworks is $\dot{\alpha}/\alpha \sim M_{PL} \sim 10^{43}/s$, which is very much larger than what is allowed by observations. However, it is at least possible that the true variations are nonzero but very small

[e]The effects of the running of h_a are described in.[2]

[f]It was argued in[28] that a variation in α would upset the fine-tuned cancellations of radiative corrections to the cosmological constant with other contributions, with enormous effect. We take the view that such arguments are not conclusive given our lack of understanding of why Λ_{cosm} is so small.

[g]We ignore possible variations in g_p because it is well described in the constituent quark model, where it is a Clebsch-Gordan coefficient.

for some reason, just as the vacuum energy is much smaller than the natural scale of $M_{\rm PL}{}^4$.

- Webb et al.[1] have reported a positive result (8), corresponding to $\dot{\alpha}/\alpha \sim 10^{-15} yr^{-1} \sim 10^{-66} M_{\rm PL}$ for constant $\dot{\alpha}/\alpha$.
- If α varies, then it is possible that other fundamental quantities, such as the other gauge couplings, Yukawa couplings, or the dimensionless ratios of the electroweak, unification, and gravity scales also vary in a correlated way that depends on the underlying physics. Such variations should be allowed for in analyzing experimental/observational results, and can in principle be a significant probe of the underlying physics.
- The comparison between different classes of observations depends on the time dependence of $\dot{\alpha}/\alpha$, which in turn depends on the type of scalar fields involved and their potentials.
- There may be long-ranged forces associated with the time variation.[14]

Acknowledgments

It is a pleasure to thank the conference organizers for support. Supported in part by Department of Energy grant DOE-EY-76-02-3071.

References

1. J. K. Webb et al., Phys. Rev. Lett. **87**, 091301 (2001); J. K. Webb, M. T. Murphy, V. V. Flambaum and S. J. Curran, Astrophys. J. Supp. **283**, 565, 577 (2003).
2. P. Langacker, G. Segre and M. J. Strassler, Phys. Lett. B **528**, 121 (2002).
3. P. A. M. Dirac, Nature **139**, 323 (1937).
4. Time variation has been discussed by many authors in many theoretical contexts. A small sample includes P. Forgacs and Z. Horvath, Gen. Rel. Grav. **10**, 931 (1979), **11**, 205 (1979); W. J. Marciano, Phys. Rev. Lett. **52**, 489 (1984), V. V. Dixit and M. Sher, Phys. Rev. D **37**, 1097 (1988); C. T. Hill and G. G. Ross, Nucl. Phys. B **311**, 253 (1988); C. T. Hill, P. J. Steinhardt and M. S. Turner, Phys. Lett. B **252**, 343 (1990).
5. For recent reviews, see T. Chiba, gr-qc/0110118; J. P. Uzan, hep-ph/0205340.
6. For constraints motivated by the need to maintain "anthropic" values of parameters,[7] see V. Agrawal, S. M. Barr, J. F. Donoghue and D. Seckel, Phys. Rev. Lett. **80**, 1822 (1998), Phys. Rev. D **57**, 5480 (1998); M. Livio, D. Hollowell, A. Weiss and J. W. Truran, Nature **340**, 281 (1989); A. Csoto, H. Oberhummer and H. Schlattl, Nucl. Phys. A **688**, 560 (2001); H. Oberhummer, R. Pichler and A. Csoto, nucl-th/9810057; T. Jeltema and M. Sher, Phys. Rev D **61**, 017301 (2000). M. Dine, Y. Nir, G. Raz and T. Volansky, Phys. Rev. D **67**, 015009 (2003);
7. For recent reviews of the anthropic principle, see C. J. Hogan, Rev. Mod. Phys. **72**, 1149 (200); B. Müller, astro-ph/0108259.

8. See, for example, J. D. Bekenstein, Phys. Rev. D **25**, 1527 (1982); T. Damour and A. M. Polyakov, Nucl. Phys. B **423**, 532 (1994); B. A. Campbell and K. A. Olive, Phys. Lett. B **345**, 429 (1995); E. Witten, hep-ph/0002297; K. A. Olive and M. Pospelov, Phys. Rev. D **65**, 085044 (2002); F. Paccetti Correia, M. G. Schmidt and Z. Tavartkiladze, hep-ph/0211122.
9. For possible space variations, see J. D. Barrow and C. O'Toole, astro-ph/9904116.
10. For applications of quintessence or other fields to time varying couplings, see for example T. Chiba and K. Kohri, Prog. Theor. Phys. **107**, 631 (2002); C. Wetterich, hep-ph/0203266, hep-ph/0301261, hep-ph/0302116; J. D. Barrow, J. Magueijo and H. B. Sandvik, Phys. Rev. D **66**, 043515 (2002), Phys. Lett. B **541**, 201 (2002). The possibility of time variation being associated with a phase transition is considered in Z. Chacko, C. Grojean and M. Perelstein, hep-ph/0204142.
11. J. D. Bekenstein, in.[8]
12. J. D. Bekenstein, astro-ph/0301566.
13. A. Kostelecky, R. Lehnert and M. Perry, astro-ph/0212003.
14. J. D. Bekenstein in [8]; K. A. Olive and M. Pospelov in [8]; G. R. Dvali and M. Zaldarriaga, Phys. Rev. Lett. **88**, 091303 (2002); J. Magueijo, J. D. Barrow and H. B. Sandvik, Phys. Lett. B **549**, 284 (2002). For a contrary view, see J. D. Bekenstein, Phys. Rev. D **66**, 123514 (2002).
15. X. Calmet and H. Fritzsch, Eur. Phys. J. C **24**, 639 (2002); Phys. Lett. B **540**, 173 (2002).
16. T. Dent and M. Fairbairn, Nucl. Phys. B **653**, 256 (2003); V. V. Flambaum and E. V. Shuryak, Phys. Rev. D **65**, 103503 (2002); V. F. Dmitriev and V. V. Flambaum, Phys. Rev. D **67**, 063513 (2003); M. Dine, Y. Nir, G. Raz and T. Volansky, Phys. Rev. D **67**, 015009 (2003); H. Oberhummer, A. Csoto, M. Fairbairn, H. Schlattl and M. M. Sharma, astro-ph/0210459; S. R. Beane and M. J. Savage, Nucl. Phys. A **713**, 148 (2003).
17. M. J. Duff, L. B. Okun and G. Veneziano, JHEP **0203**, 023 (2002).
18. J. N. Bahcall, C. L. Steinhardt and D. Schlegel, arXiv:astro-ph/0301507.
19. L. L. Cowie and A. Songaila, Astrophys. J. **453**, 596 (1995).
20. A. Y. Potekhin *et al.*, Astrophys. J. **505**, 523 (1998).
21. J. D. Prestage, R. L. Tjoelker and L. Maleki, Phys. Rev. Lett. **74**, 3511 (1995).
22. Y. Sortais et al., Physica Scripta **T95**, 50 (2001).
23. T. Damour and F. Dyson, Nucl. Phys. B **480**, 37 (1996).
24. Y. Fujii *et al.*, Nucl. Phys. B **573**, 377 (2000).
25. E. W. Kolb, M. J. Perry and T. P. Walker, Phys. Rev. D **33**, 869 (1986); L. Bergstrom, S. Iguri and H. Rubinstein, Phys. Rev. D **60**, 045005 (1999); P. P. Avelino *et al.*, Phys. Rev. D **64**, 103505 (2001); J. J. Yoo and R. J. Scherrer, Phys. Rev. D **67**, 043517 (2003).
26. S. Hannestad, Phys. Rev. D **60**, 023515 (1999); M. Kaplinghat, R. J. Scherrer and M. S. Turner, Phys. Rev. D **60**, 023516 (1999); C. J. Martins, A. Melchiorri, G. Rocha, R. Trotta, P. P. Avelino and P. Viana, astro-ph/0302295.

27. See, for example, P. Langacker and N. Polonsky, Phys. Rev. D **52**, 3081 (1995).
28. T. Banks, M. Dine and M. R. Douglas, Phys. Rev. Lett. **88**, 131301 (2002).

Maurice Goldhaber

Proceedings of the Dirac Centennial Symposium
Howard Baer and Alexander Belyaev

Amending the Standard Model of Particle Physics

Maurice Goldhaber
Physics Department
Brookhaven National Laboratory
Upton, NY 11973

Some of my earlier arguments, suggesting modifications of the Standard Model of Particle Physics (see ref. [1]), are elaborated and extended. Rules deduced from the known properties of elementary fermions are sharpened and extended in the first part. Conclusions drawn from the rules in the second part are also honed and expanded and an estimate of the neutrino mass eigenstates is added. In the third part, a tentative explanation of the rules is discussed. In my earlier paper, I suggested replacing the *point-sources* postulated by the Standard Model for each generation by *finite 'source-shapes'*, equal for all elementary fermions of a generation *and systematically* decreasing in volume from the first to the third generation, thus increasing the effect of self-interactions. According to the rules a correlation exists between the mass of an elementary fermion and the strength of its self-interaction, thus an increase in self-interactions would resolve the problem of the hierarchical masses. A possible connection between the existence of only three generations and the three-dimensionality of space also is discussed. In the epilogue the question is explored whether finite source-shapes for the elementary fermions can be reconciled with fundamental theoretical tenets.

Part 1

To set the stage, let us briefly summarize some of the empirical facts from which the rules are deduced. A century of research has established the existence of a 'periodic system' of *only* three generations of spin $\frac{1}{2}$ elementary fermions, labeled by a generation number i (1-3). As shown in Table I, each generation consists of four *types* of elementary fermions: *two leptons*, one of charge −1 (e, μ, τ), with its 'own' neutrino of charge 0 (ν_e, ν_μ, ν_τ), and *two quarks*, one of charge +2/3 (u, c, t), associated with a specific quark of charge -1/3 (d, s, b), by the order of their discovery, later justified by a *physical affiliation* (see below).

For general arguments, we shall refer to the ith generation as

Table I
The Three Generations of Elementary Fermions

$$\begin{pmatrix} t & \tau \\ b & \nu_\tau \end{pmatrix} \quad i = 3$$

$$\begin{pmatrix} c & \mu \\ s & \nu_\mu \end{pmatrix} \quad i = 2$$

$$\begin{pmatrix} u & e \\ d & \nu_e \end{pmatrix} \quad i = 1$$

$$\begin{pmatrix} u_i & e_i \\ d_i & \nu_i \end{pmatrix}$$

and call elementary fermions of the same type a 'family', referred to as f_i (comprising u_i, d_i, e_i and ν_i.)

The Standard Model of Particle Physics (SM)[2] postulates chiral symmetries for the interactions of elementary fermions. Since such symmetries lead to zero masses for the elementary fermions, their actual masses are attributed to symmetry breaking by the Higgs mechanism. Such a mechanism was needed to give masses to the gauge bosons, $W^{\pm}$ and Z^0, but it can only accommodate masses for the elementary fermions, not predict them.

Each family of elementary fermions possesses some or all of the four known universal interactions that are believed to be 'elementary' (see Table II), while the interactions, e.g., of complex hadrons built of quarks, are derivative.

Table II
Universal Interactions of the Elementary Fermions

Interactions	Approximate Relative Strength	Leptons ν_i	Leptons e_i	Quarks u_i d_i
Strong	1			X
Electro-magnetic	10^{-2}		X	X
Weak	10^{-5}	X	X	X
Gravitational	10^{-39}	X	X	X

The interactions, shown in hierarchical order of their strength, are found to vary with energy ('running constants'), characterized here at ~1GeV.

The MeV equivalents of the *mean* mass, deduced from the known ranges for the masses of elementary fermions[3] are plotted in Figure 1. Ever since the mass of the neutron was found to be heavier than that of the proton,[4] it seemed puzzling why for a nucleon doublet, with presumably equal nuclear interactions, the mass difference is not the reverse, as one might expect due to the proton's Coulomb interaction. This mass difference and those found for other hadron multiplets were later ascribed to a difference between the masses of the d and u quarks of which they are built. The mass difference and the mass ratio have been calculated for these quarks: Narison[5] found that $m_d - m_u \simeq 3$ MeV, and the most recent lattice calculation by Nelson, Fleming and Kilcup[6] yielded $m_u/m_d = 0.410 \pm 0.036$. From these two values we can deduce $m_d \simeq 5$ MeV and $m_u \simeq 2$ MeV. The puzzle thus has shifted: In spite of the fact that the absolute charge of the u quark being twice that of the d quark, the u quark is the *lighter* one! In Part 3, I discuss a possible explanation for this puzzle.

Approximate values for the neutrino mass eigenstates are deduced below.

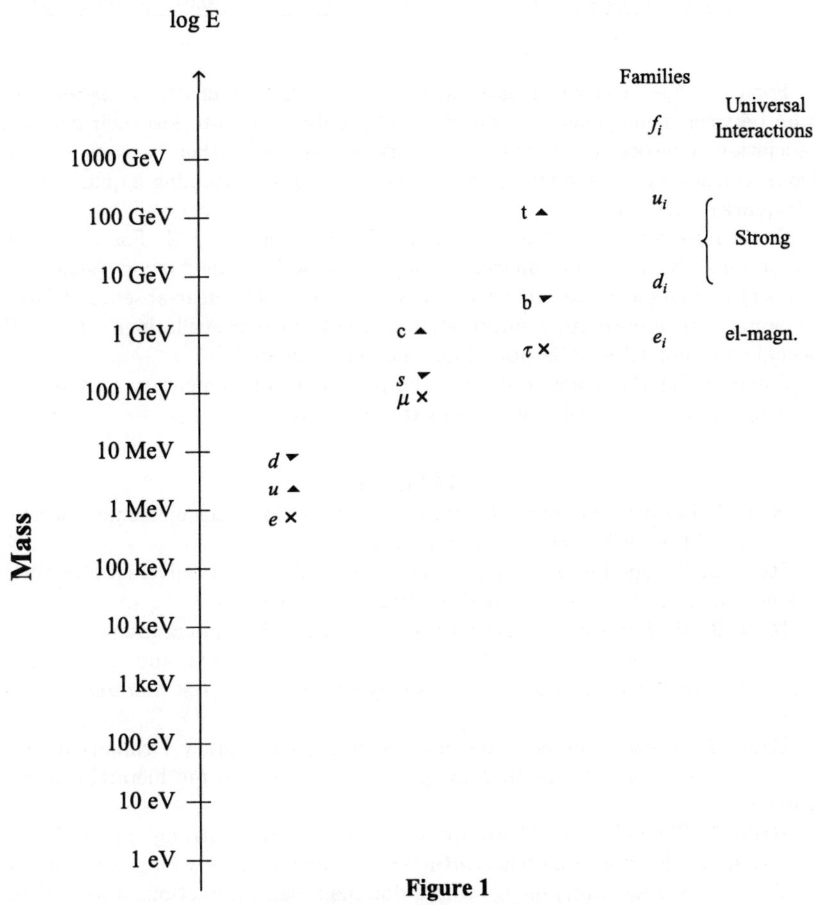

Figure 1

The mean of the masses of the elementary fermions

For quarks the so-called 'current' masses are given, obtained from lattice calculations at ~2 GeV, except for the directly measured mass of the top quark (3). The *u* and *d*, masses are deduced from their calculated difference and their ratio.

The heavier elementary fermions decay by weak interactions into lighter ones. For quarks, this process is described by the unitary C-K-M matrix – named for Cabbibo, Kobayashi and Maskawa[7] – that assumes 'weak mixing' of quarks.

The mean values of the matrix elements[8] are found to depend on the absolute 'generation distance' $|i-j|$, as shown in Table III.

Table III

The Mean Values of the Absolute CKM Matrix Elements $|\overline{V}_{u^i d^j}|$

$\|i-j\|$	0			1				2	
i	1	2	3	1	2	2	3	1	3
j	1	2	3	2	1	3	2	3	1
	0.9749	0.9745	0.9992	0.223	0.223	0.041	0.041	0.004	0.009

For $i=j$, the matrix elements have near unit value, indicating a *special affiliation* between same-generation quarks, and justifying à posteriori their historical association. Quarks are confined to 'colorless' hadrons, either baryons (e.g. nucleons, containing three quarks) or mesons (e.g. pions, containing a quark and an anti-quark) – see ref. [2]

Some noteworthy conclusions can be drawn from the SM. For the known interactions, the number of quarks and leptons as well as the "flavor" (generation number) of charged leptons is very nearly conserved. The near-absence of flavor-changing neutral currents is interpreted as due to near cancellations – the 'GIM mechanism', named for Glashow, Iliopoulos and Maiani.[9]

Some of the *Rules* deduced from the properties of elementary fermions – a few long known – are elaborated and extended here.

Rules

Rule 1. Besides its dominant (strongest) interaction, each elementary fermion possesses *all* the known weaker interactions.

Rule 2. Except for the gravitational interaction, *each* universal interaction is *dominant* in at least one 'family' of elementary fermions.

Rule 3. *Within* each generation the mass of an elementary fermion is found to be correlated with the strength of *its* dominant interaction, and thus with the hierarchical universal interactions. This may be called the *first hierarchical mass relation.*

Rule 4. In each family f_i the masses of the elementary fermions increase as i increases – a *second hierarchical mass relation*, called *the* hierarchical mass problem.

Rule 5. Though the SM attributes *equal* dominant strong interactions to the $u_i$and d_i, the mass difference $m(u_i)-m(d_i)$ increases as i increases, paralleling the situation for $m(e_i)-m(v_i)$ whose dominant self-interactions have *different* strengths.

Rule 6. A change from one quark to another, induced by the weak interaction, is represented by the C-K-M matrix whose elements decrease as the generation difference increases – an '*anti-hierarchical*' relation, with $|\overline{V}_{u^i d^j}| \simeq \overline{V}_{u^j d^i}|$.

Rules, sometimes called regularities, laws, systems, principles etc., formulated in many fields of science, can often be considered as *proto-theories* because interpolations and extrapolations from them have proven to have predictive power, leading to important discoveries. They hint at the existence of *hidden* causes, and any exceptions from rules may indicate the existence of more than one such cause. Successful guesses at such hidden causes had often led to revolutionary theoretical developments.

Though the SM is remarkably successful in many respects, it leaves a gap in understanding most of the above Rules. Through the Rules *nature speaks to us*, and it behooves us to listen. Extrapolations from the Rules suggest additions and modifications for the SM, and since most of the Rules are only qualitative, these extrapolations usually will be only qualitative.

Part 2

Conclusions suggested by the Rules

(1) The Role of Self-Interactions. Rule 3 suggests that dominant self-interactions play an important role in the resulting masses of the elementary fermions, and since the SM cannot predict actual masses, this Rule is a helpful qualitative guide.

(2) Do Undetected Elementary Fermions Exist? Rule 2 suggests the *possibility* that a *fifth member* of *each* generation exists, an elementary fermion where the gravitational interaction *is* dominant and, in the absence of still weaker interactions, the *only one.* Let us call such hypothetical elementary fermions provisionally *gravi-fermions,* and refer to them as $g_i(1/2)$, to differentiate them from the zero mass graviton g(2). According to Rule 3 the $g_i(1/2)$ would be expected to have exceedingly small masses, and because of the weakness of the gravitational interaction, cannot be detected directly at energies much lower than the Planck mass. But if they do exist, they might connect general relativity and quantum mechanics in a new way.

(3) Do the Three Generations differ in a Hidden Physical Property? Rules 4 and 6 imply that elementary fermions 'know' to which generation they belong. This suggests that the generation number i is not just a label, but stands for a hidden *physical property* that changes *progressively* from the first to the third generation in a way that enhances the effect of self-interactions, thus leading to hierarchical masses. Unlike Mendeleev's periodic system of the chemical elements, for which only practical limits are known, the 'periodic system' of the elementary fermions stops at three generations. In Part 3 we discuss a possible interpretation of the hidden physical property, and why we have three generations only.

It is worth reminding oneself that Pauli's 'closer look' at spectroscopic data led him to suggest a double-rule, known as the Exclusion Principle, that expressed the two previously hidden properties of existence of

1.) A classically not describable 'two-valueness', equivalent to a new quantum member, and

2.) that no two electrons can have identical four quantum numbers.
The two aspects of the rule were later explained

a.) by the suggestion of the electron spin by Uhlenbeck and Goudsmit and

b.) by the anti-symmetry of fermion wave functions, leading to Fermi-Dirac statistics.

(4) Does a New Universal Interaction Exist? The apparent contradiction between Rules 3 and 5 suggests the possibility that the u_i, but not the d_i, possess, besides their strong interaction, a new dominant *hyper-strong* interaction, increasing in importance as i increases, for which a possible reason is discussed in ref. [1]
Recently the Belle Collaboration at KEK reported a large deviation from the SM: *Observation of double $c\bar{c}$ production in e^+e^- annihilation at $\sqrt{s} \simeq 10.6$ GeV.*[10] They find the probability of this process to be about an order-of-magnitude larger than expected from the SM. This may hint at the possibility that a hyper-strong interaction may affect the production cross section of c quark, a possibility that Dmitri Kharzeev and I are studying.

If we tentatively accept the conclusions (2) and (4), Table IV would replace Table II.

Table IV
Universal Interactions of the Elementary Fermions

Interactions	Approximate Relative Strength	Gravi-Fermion	Leptons ν_i	Leptons e_i	Quarks d_i	Quarks u_i
(Hyper-Strong	>1					X
Strong	1				X	X
Electro-magnetic	10^{-2}			X	X	X
Weak	10^{-5}		X	X	X	X
Gravitational	10^{-39}	(X)	X	X	X	X

(5) Approximate Evaluation of the Neutrino Mass Eigenstates[a] .
The Super-Kamiokande collaboration[11] established the existence of oscillations for atmospheric μ-neutrinos, and the SNO collaboration[12] obtained direct evidence for oscillations of the solar e-neutrinos, in agreement with the earlier indirect evidence of J. Bahcall, R. Davis and coworkers.

Oscillations prove that neutrinos have finite masses and that their flavor (generation number) is not conserved. Finite neutrino masses are still often called a

[a] An earlier version of this section was reported at the Stony Brook Conference on **"Neutrinos and Implications for Physics Beyond the Standard Model"**, Stony Brook, New York, October 11-31, 2002.
See http://www.insti/physics.sunysb.edu/ips/conf/neutrino/talks/goldhaber.pdf

mystery because the SM predicts zero masses. Rule 3, however, suggests that neutrinos, with their weak dominant interaction, should have finite, though small masses.

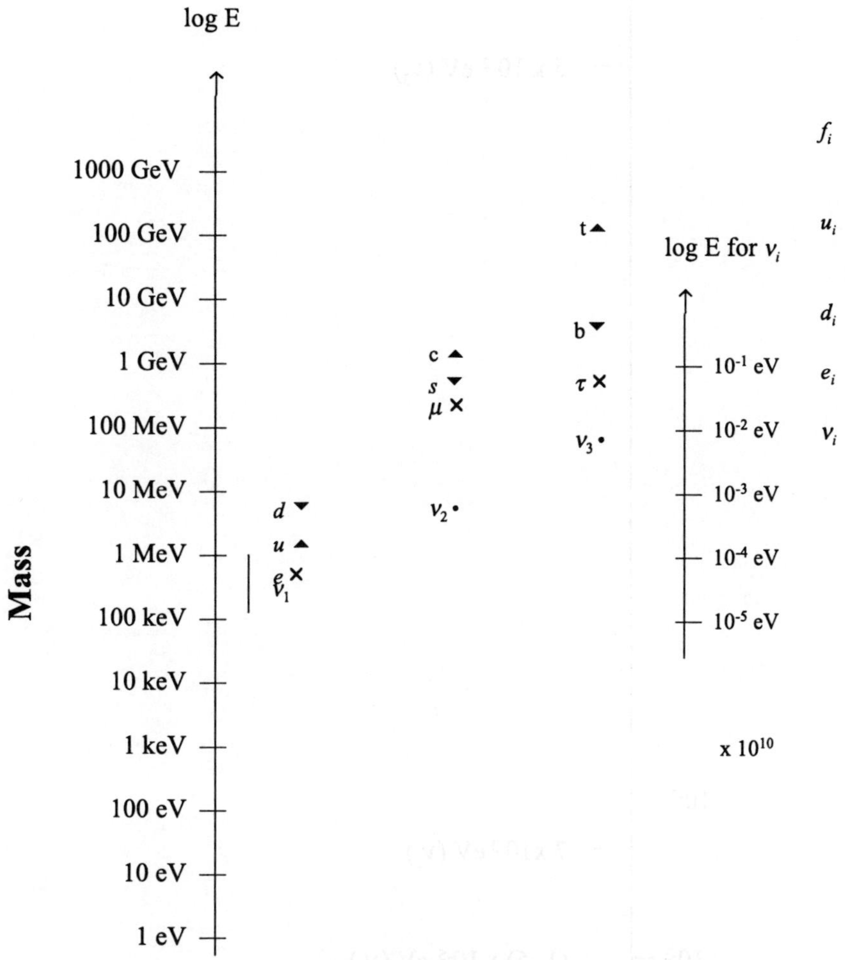

Figure 2

The mean masses of the elementary fermions (left scale) and of the approximate neutrino mass eigenstates - moved up by a factor of 10^{10} (right scale). The line to the left of the symbol ν_1 indicates the uncertainty in its proposed mass.

Figure 3

The estimated values of the neutrino mass eigenstates.
The thickness of the line for ν_1 symbolizes the estimated uncertainty in its mass value.

It has often been pointed out that oscillation experiments, yielding values for Δm^2 between different neutrino mass eigenstates do not allow us to distinguish - without further assumptions - between three possible scenarios: hierarchical, anti-hierarchical, or nearly degenerate neutrino mass eigenstates.

Inspection of the masses of the elementary fermions with equal dominant interactions that led to Rule 4 (see Fig. 1) yields the following sub–rule:

$$m_{fi}^2 \ll m_{fi+1}^2$$

Assuming that these rules also hold for the neutrino mass eigenstates m_i the following relations should hold:

$m_1 < m_2 < m_3$
and
$m_1^2 \ll m_2^2 \ll m_3^2$.

Since no evidence for oscillations of atmospheric μ-neutrinos into e-neutrinos above the background of atmospheric e-neutrinos has so far been detected, two-flavor oscillations, $\nu_\mu \to \nu_\tau$, are considered to be a good approximation.

From the measured survival probability
$p(\nu_\mu \to \nu_\tau) = 1 - sin^2 2\theta sin^2(1.27\Delta_{32}m^2[\mathrm{eV}^2]\mathrm{L[km]/E}_\nu[\mathrm{GeV}])$
one obtains[11]

$$\Delta_{32}(m^2) \equiv m_3^2 - m_2^2 \approx 2.5 \times 10^{-3}(\mathrm{eV})^2.$$

From the preferred LMA MSW solution for solar neutrino oscillations, one obtains[12]

$$\Delta_{21}(m^2) \equiv m_2^2 - m_1^2 \approx 5 \times 10^{-5}(eV)^2.$$

Neglecting m_i^2 relative to m_{i+1}^2, we find

$$m_3 \approx \sqrt{\Delta_{32}} \approx 5 \times 10^{-2} \text{ eV}$$

and

$$m_2 \approx \sqrt{\Delta_{21}} \approx 7 \times 10^{-3} \text{ eV}.$$

The most accurately known f_i masses are those of the charged leptons and it is of interest to compare the ratios $\mathrm{m}_i/\mathrm{m}_{i+}$ with the ratios $\mathrm{m}_{ei}/\mathrm{m}_{ei+1}$ that are intermediate between the ratios for the u_i and d_i families. We find that $\mathrm{m}_2/\mathrm{m}_3 \approx 1.4\mathrm{x}10^{-1}$, which is ~2.4 times larger than the ratio

$$m_\mu/m_\tau = \frac{105.66}{1.777 \times 10^3} = 5.9 \times 10^{-2}.$$

Assuming, as an approximate guide, that the masses of the ν_i and e_i families are nearly parallel on a log scale (see Fig. 1), one obtains

$$m_1/m_2 \approx m_e/m_\tau = 4.8 \times 10^{-3}.$$

For an estimate of the uncertainty of the m_1 value we assume for m_1/m_2 a similar deviation by the factor of ~ 2.4 found for m_2/m_3. In this case, however, it may be safer to assume that the deviation could be in either direction, yielding a range for

$$m_2 \approx (1-5) \times 10^{-5} \text{ eV}.$$

As a further test we deduce m_1 differently by using the relation

$$m1 = m_e/m_\tau \cdot m_3 = \frac{0.51}{1.777 \times 10^3} \cdot 5 \times 10^{-2} \text{ eV} = 1.4 \times 10^{-5} \text{ eV},$$

which falls within the suggested range.

The KamLAND Collaboration[13] finds a result compatible with the LMA MSW solution, preferred by SNO. Such a solution would independently suggest a hierarchical order of the neutrino mass eigenstates involved: $m_2 > m_1$.

The estimated neutrino mass eigenstates are compared in Figure 2 with the masses of the other elementary fermions, given in Figure 1, and are shown in Figure 3 on a linear energy scale. The masses obtained for the neutrino mass eigenstates are $10^{11\pm1}$ times smaller than those of the corresponding charged leptons.

Part 3

Attempts to Explain the Rules

To explain the origin of the rules I made several speculative assumptions,[1] speaking, so to say, *to* nature, without knowing whether *it* listens! Conclusions reached in this part may contradict some well-developed fundamental theories, as discussed below.

Starting with Conclusion 3, I asked what kind of hidden physical property might change from the first to the third generation in a way that would lead to hierarchical masses? And, why does nature stop repeating itself after three generations? While retaining one of the SM's two assumptions, the well established universality of interactions, we tentatively hypothesized that the hidden physical property implies replacing the other assumption, of equal *point sources* for each generation by postulating *finite source-shapes*, equal for all elementary fermions of a generation, but with volumes *shrinking* systematically from the first to the third generation. This would increase their 'singularity' as i increases, elevating the effect of self-interactions, and thus leading to hierarchical masses.

A physical point is a theoretical extrapolation, while everything we observe *directly* has three dimensions. However, when a body extends much less in some of the three spatial directions than in the remaining ones, we tend to assign to it the *dominant* dimension, e.g. that of a sphere (or spherical shell), a disk or a rod, with finite thicknesses and no sharp edges. If the universal interactions had such source-shapes, they would naturally connect the existence of only three generations with the three-dimensionality of space: as i increases from one to three, the dominant dimensionality of the generations would decrease from three to one.

The Sign of the mass difference $d - u$

Although the value of the masses of the elementary fermions are correlated with the strength of their dominant self-interaction, subdominant self-interactions also must play some role, as they *do* when *they* are the dominant interactions in lighter elementary fermions. Following Dirac's example, we consider self-interactions classically.[14] (See a further discussion below.) Depending on whether the subdominant self-interaction has the same or opposite sign of that of the dominant one, the resulting *total* self-interaction would correspondingly increase or decrease. The combination of an attractive strong interaction and a repulsive electro-magnetic one would reduce the expected self-interaction for *both* u and d, independent of the sign of their charge, but because of the larger electric charge of u, its mass would be *more* depressed than that of d. For i=1, the effect of a hyperstrong interaction, rising rapidly as i increases (see ref [1] is apparently not sufficient to lift the mass of u above that of d. But, in spite of the Coulomb depression expected to affect all u_i masses, the hyperstrong interaction wins out for i=2 and 3.

Epilogue

It is ironic that the *zero* masses of the elementary fermions, predicted by the SM, are postulated to be changed by the Higgs mechanism into the observed masses, which can accommodate but not deduce the masses, while half a century ago the *infinite* masses – implied by point-charges – were renormalized into the *same* masses! Though it is widely considered that masses are secondary quantities, the possibility of calculating them is relinquished by the SM. With finite source-shapes renormalization might not be needed, as Dirac suspected. He proposed to consider the electron classically as a charged conducting surface, with a surface tension to prevent it from flying apart under the repulsive forces of the charge.

For many of its quantitative predictions the SM must explicitly use *empirically* obtained quantities. When dealing with interactions of elementary fermions at wavelengths that are larger than the sizes of the source-shapes, point sources may remain a good approximation. Thus, appreciable deviations from the SM might be detectable only at rather high energies, though small deviations from its predictions, reported for some precision measurements, might conceivably be connected with effects of finite source-shapes.

While the C-K-M matrix elements have been commonly anti-correlated with the hierarchical masses, presumably secondary quantities, we suggested instead to correlate them with the overlap of the source-shapes (see ref. [1]).

With finite source-shapes, the possibility of excited states of elementary fermions exists. Though such shapes have finite size in common with multi-dimensional string theories, our approach is clearly very different.

I said in ref. 1 that should it be possible to change the qualitative suggestions into quantitative ones without contradicting relativistic quantum mechanics, the number of independent parameters needed in the SM might be considerably reduced, depending on the number of parameters needed to describe the source-shapes.

Some might say that I have only 'deconstructed' what can be ascribed to the Higgs mechanism (if one only knew how to calculate its influence). Even if this were so, my considerations might prove to be of heuristic value.

A few past rules have been only partially successful, with exceptions being early indications of the existence of more than one cause for the rules. In spite of its remarkable prediction of new elements by interpolations, Mendeleev's system had exceptions. These were explained after the hierarchical atomic weights, by which he was guided, were replaced by the integer atomic numbers, equal to the number of protons in atomic nuclei. It is interesting that Mendeleev, in a Faraday lecture delivered in 1889 twenty years after he developed the periodic system, speculated that it would need a new 'chemical mechanics' to understand it. Wheeler,[15] who drew attention to this lecture, calls this a premonition of the quantum, a decade ahead of Planck!

LoSecco[16] concludes from an analysis of the neutrino signals from the Supernova SN1987A, recorded by Kamiokande and IMB, that neutrinos and anti-neutrinos are equally affected by the intervening gravitational fields. Direct attempts to 'weigh' anti-particles (anti-hydrogen) are in progress at CERN (see Gabrielse et al,[17]).

What has been called elementary or fundamental has varied with time. First atoms were considered elementary, until they were shown to be made up of 'elementary' nuclei and electrons, then nuclei, in turn, were shown to be composed of 'elementary' nucleons (protons and neutrons). Finally, nucleons were shown to consist of elementary quarks (which can, however, not be studied as individual free particles). Thus, a particle was considered elementary as long as it was not found to contain distinct sub-units, a definition that would include finite source-shapes for the time being. Even when it was already known that both nuclei and nucleons, have a finite extent, they were treated as elementary (effective field theory). In his Nobel lecture, t'Hooft[18] however, concludes "... if R [the radius of a particle] were finite it would be difficult to take into account that *forces* acting on the particles must be transmitted by a speed less than that of light, as is demanded by Einstein's theory of special relativity. If the particle were deformable, it would not be truly elementary. Therefore, finite-size particles cannot serve as a good basis for a theory of elementary objects."

Some theories were developed in intermediate steps, often consisting of radical changes *grafted* on existing theories, and in spite of provisional successes, ultimately superseded by new formulations. The evolution of quantum mechanics provides an instructive example. The remarkable success of Bohr orbits, and what might be called 'de Broglie orbits', grafted onto classical mechanics, was later understood when the genesis of a detailed theory, wave mechanics, proved that these orbits approximately coincided with the mean of their wavefunctions. Schrödinger thought that his wavefunction indicated that the electron is spread out, but finally Born interpreted the absolute square of the wavefunction as the probability of finding an electron. Thus, earlier theories, though they did not survive, caught important aspects, 'as if' they were correct, though earlier versions often give sufficiently good results with simpler calculations e.g., for spectroscopicsts and chemists. As we have seen, Dirac went back to a classical theory of the electron when he attempted to calculate its mass.

What Wilczek[19] has called the Einstein-Wheeler dream, 'mass without mass', he found nearly fulfilled for the visible Universe because its nucleons owe their mass mainly to the internal kinetic energy of the light d and u quarks and the zero mass gluons. However, this would not be so for the fleetingly existing particles made of elementary fermions of the second and third generations that have, according to the SM, considerable masses as well as energy. But what I suggest to call the Einstein-Dirac-Wheeler dream that all mass may be energy does not look like an impossible dream.

Acknowledgements

I have profited from discussions with many physicists, including those who disagree with some aspects of this paper: A.J. Baltz, T.C. Blum, M. Blume, M.J. Creutz, M.V. Diwan, F.J. Dyson, A.S. Goldhaber, T.J. Goldman, R. Jaffe, C.K. Jung, W. J. Marciano, F.E. Paige, R.E. Shrock, G.F. Sterman and A.S. Wightman.

P.S.

I realize now that Rule 1, from which no conclusions have been drawn, is trivial, because quarks have electro-magnetic besides strong interactions, and electromagnetic interactions are connected with the weak interactions by the electroweakunification.

References

1. Proc. Nat. Acad. Sc. USA **99**. 33, (2002) (With minor corrections): hep-ph/0201208.
2. P. D. B. Collins, A. D. Martin & E. J. Squires, (1989) *Particle Physics and Cosmology* (Wiley, New York); D. H. Perkins, (2000) *Introduction to High Energy Physics* (Cambridge Univ. Press, Cambridge, U.K.); Donahoe et al., G. t'Hooft, (1997) *In Search of the Ultimate Building Blocks* (Cambridge Univ. Press, Cambridge, U.K.)
3. Particle Data Group, Phys. Rev. D **66**, July 2002, 01000125 – 129
4. J. Chadwick and M. Goldhaber, Nature **134**, 237 (1934)
5. S. Narison, Nucl. Phys. (Proc. Suppl.) B86, 242 (2000) or hep-ph/0202200 (chapter of forthcoming book on QCD for Cambridge Univ. Press) and private communication.
6. D.R. Nelson, G.T. Fleming and G.W. Kilcup, Phys. Rev. Lett. 90,
7. N. Cabbibo, Phys. Rev. Lett. **10**, 531-533(1963); M. Kobayashi & K. Maskawa, (1973) Prog. Theor. Phys. **49**, 652-657
8. F. J. Gilman, K. Kleinknecht & B. Renk, see Ref. 3, p 010001-113
9. S. L. Glashow, J. Iliopolulos & L. Maiani, (1970) Phys. Rev. D 2, 1285-1292
10. Belle Collaboration, K. Abe, et al, Phys. Rev. Lett. **89**, 142001 (2002)
11. Super Kamiokande Collaboration, see T. Kajita and Y. Totsuka, Rev. Mod. Phys. 73, (2001) 83-118, and C.K. Jung, T. Kajita, T. Mann and C. McGrew, Ann. Rev. Nucl. Part. Sci. 2001, 451.
12. SNO Collaboration, Q.R. Ahmad et al., Phys. Rev. Lett. **89**, 011301 (2002)
13. KamLAND Collaboration, R. Eguchi et al., Phys. Rev. Lett. **90**, 021802 (2003)

14. P.A.M. Dirac, Proc. Roy. Soc. (A) **268**, 57 (1962)
15. J.A. Wheeler, Transection of the New York Academy of Sciences, Series II, **233**, 745 (1971)
16. J.M. LoSecco, Phys. Rev. D **39**, 1013-19 (1989)
17. G. Gabrielse et al, Phys. Lett. B **507**, 1-6 (2001)
18. G. t'Hooft, Rev. Mod. Phys. **72,** 333, see p. 334 (2000)
19. 19. F. Wilczek, Phys. Today **11**, November 1999, and **13**, January 2000